# Praise for Sacrifice

"Dr. Jim Thorp stood as a rock-solid pier of truth and integrity in a sea of corruption that swept into the field of obstetrics and gynecology. Thorp stood tall, called out corruption and malfeasance, and was willing to pay any price to protect mother and baby from this attack. This book should be shared from mother to obstetrician, one-by-one, all over the globe. Thorp is the rock!"

—**Peter A. McCullough, MD, MPH, author of**
*The Courage to Face COVID-19*

"The only writer I would trust to carry a reader through a book with information as shattering as that in *Sacrifice* is Celia Farber. A veteran of our long, long war, and a dear friend, she can be trusted to ensure that the human stories are never eclipsed by the hard data. A great, great work, *Sacrifice* is certain to make a lasting impact."

—**Vera Sharav, Holocaust survivor, human rights activist, and founder of the Alliance for Human Research Protection**

"What if it turned out that the world you thought you were living within is surrounded by smoke and mirrors to distract you from the evil that increasingly encroaches on all of us? What if you opened your eyes—and you finally saw this evil? If taken as seriously as it should and must be, *Sacrifice*, by James Thorp and Celia Farber, could become an enormous awakening for many readers—pulling away the curtain on major events going on in the world that involve mass murder and yet are being entirely ignored and suppressed by the major media and all the big-name influencers and politicians."

—**Dr. Peter Breggin, psychiatrist, psychotherapist, and author of *COVID-19 and the Global Predators: We Are the Prey***

"Readers may or may not agree with Dr. Thorp's conclusions regarding the malign intent of those who engineered the miserably failed response to COVID. But after reading chapter one, it will be impossible not to understand how Dr. Thorp came to them. The authors are well aware of the scorn and vilification that will be their reward for writing this book. I hope everyone will reward their courage by reading *Sacrifice* with an open mind that will allow their eyes to be opened."

—**Ron Johnson, US Senator, Wisconsin**

"Dr. James Thorp has long been the only, and now, among the very few, obstetricians to come forward and speak up about the damage that 'the deadliest vaccine in history' has done to babies in utero, to moms-to-be, and to newborns. He also exposed the industrial-scale corruption of the field of obstetrics by vaccine money. If history is ever properly written, Dr. Thorp will be honored as the man who sought to save generations unborn."

—**Naomi Wolf, PhD, author of *The Bodies of Others*, cofounder and CEO of DailyClout.io**

"The authors provide a personal first-hand, referenced, and accurate, exposé of the mRNA 'vaccine' debacle promoted by the dysfunctional leadership of ACOG, ABOG, and the SMFM. By early 2021, it was clear that not one pregnant woman should have ever been given a COVID-19 'vaccine,' which are still in use. Every mother in the US should be outraged, yet the silence is deafening due to medical censorship."

—**Dr. Steven J. Hatfill, medical advisor, 2020 Executive Office of the President**

"Dr. James Thorp builds a strong case for the evolution of a COVID catastrophe. Brutal techniques of censorship and propaganda, he suggests, were deployed in a manner applied during similar medical disasters in our past. Calling upon his clinical experience and the historical record Dr. Thorp tells a grim tale of betrayal and corruption."

—**Drew Pinsky, internist and addiction medicine specialist, television host, author, and public speaker**

"A True Hero amongst us, Dr. James Thorp speaks against the 'Covid Vaccine' calling it the 'Highest Kill Rate in History.' His brave words saved countless lives. Thank you for being a Guardian Angel to the vulnerable."

—**Alex Jones**

"*Sacrifice: How the Deadliest Vaccine in History Targeted the Most Vulnerable* shines light on the devastating harms the COVID shots have had on our unborn. Beautifully written, the book is coauthored by one of the few obstetricians in the US with the courage to publicly speak about the unprecedented rise in fetal demise following the rollout of the COVID shots. Dr. Thorp and Celia Farber have written a powerful expose on how and why 99.99 percent of obstetricians have ignored the oath they took to first Do No Harm. I had chills within the first five minutes of reading this book—I cannot recommend highly enough."

—**Dr. Mary Bowden, founder of Americans for Health Freedom**

# SACRIFICE

# SACRIFICE

## HOW THE DEADLIEST VACCINE IN HISTORY TARGETED THE MOST VULNERABLE

JAMES THORP, MD
AND CELIA FARBER

Skyhorse Publishing

Copyright © 2025 by James Thorp and Celia Farber

All Rights Reserved. No part of this book may be reproduced in any manner without the express written consent of the publisher, except in the case of brief excerpts in critical reviews or articles. All inquiries should be addressed to Skyhorse Publishing, 307 West 36th Street, 11th Floor, New York, NY 10018.

Skyhorse Publishing books may be purchased in bulk at special discounts for sales promotion, corporate gifts, fund-raising, or educational purposes. Special editions can also be created to specifications. For details, contact the Special Sales Department, Skyhorse Publishing, 307 West 36th Street, 11th Floor, New York, NY 10018 or info@skyhorsepublishing.com.

Skyhorse® and Skyhorse Publishing® are registered trademarks of Skyhorse Publishing, Inc.®, a Delaware corporation.

Visit our website at www.skyhorsepublishing.com.

Please follow our publisher Tony Lyons on Instagram @tonylyonsisuncertain.

10 9 8 7 6 5 4 3 2 1

Library of Congress Cataloging-in-Publication Data is available on file.

Hardcover ISBN: 978-1-5107-8329-4
eBook ISBN: 978-1-5107-8330-0

Cover design by Brian Peterson

Printed in the United States of America

# Dedication from Dr. James A. Thorp

To my best friend, soulmate, and beautiful wife Maggie M. Thorp, JD, MACP, thank you for being on my team, standing by me, loving me, and joining me in our fight against the evil perpetrated upon humanity by the dark, sinister, demonic forces in charge of this world.

To my mother and father, Ken and Mollie Thorp. I love you. God rest your souls, and I look forward to being with you again. Thank you for instilling in me a desire to work harder, learn more, better serve the poor, and deliver compassionate care as a physician. Thanks, Mom, for being a role model for me as a labor and delivery nurse and instilling in me a love for the Lord.

To Dr. Semmelweis, you are my hero, and I look forward to meeting you. Thank you, Dr. Semmelweis, for teaching me how to save lives as a physician, to always avoid groupthink, and to eradicate false narratives responsible for killing and injuring our patients. You taught me well. Thank you, Dr. Semmelweis, also for inspiring me to always be involved with meaningful clinical research to better serve my patients.

Fifty years ago, when I first read Morton Thompson's account of Semmelweis's *The Cry and the Covenant*, I *never* could have imagined that I would be facing an even more dire situation in this pandemic era for our obstetrical patients than that of Ignaz Philip Semmelweis in the mid-nineteenth century.

*What has been done, will be done again. There is nothing new under the sun.* (Ecclesiastes 1:9).

# Contents

A Note on the Text | A Note from Dr. Thorp — xi
Foreword by Christiane Northrup, MD — xiii
Introduction — xvi

## PART ONE

Chapter One: Elixirs of Death: A Day in the Life of an Obstetrical Doctor — 3

Chapter Two: Nurse Bearing Witness: How a Delivery Ward Became a Death Ward in 2021 — 10

Chapter Three: Inviting the Devil In: The Machine Model of Biology, Gene Therapy, and the Fall — 31

Chapter Four: Interview: A Paramedic Who Became an Enemy of the Covid State — 47

Chapter Five: How They Did It: A Roaring Tidal Wave of Money — 57

Chapter Six: Twin Towers of Catastrophe in Maternal-Fetal Medicine: Thalidomide and DES — 75

Chapter Seven: Eight Candles: The Holocaust Hits Home — 87

## PART TWO

Chapter Eight: The Obsolete Man — 97

Chapter Nine: Getting Fired: A No-Cause Termination with a Cause — 100

Chapter Ten: The Threat: ABOG's Attempt to Silence Me — 107

| Chapter Eleven: | My Response: An Open Letter from James A. Thorp, MD, to ABOG | 116 |
| --- | --- | --- |
| Chapter Twelve: | Vaccine Masquerade: The Rise of the Overnight Vaccine Cult and the Attack on Reproduction | 134 |
| Chapter Thirteen: | The Turning Tide: A Story of Hope and Perseverance | 159 |

*Epilogue*   167
*Endnotes*   169
*Index*   181

## A Note on the Text

Dr. James Thorp is responsible for the scientific and medical content, and Celia Farber for the storytelling and narrative shaping.

## A Note from Dr. Thorp

I am donating my proceeds to a tax-exempt fund at The Wellness Company, TWC.health. The Wellness Company will match all donations. This tax-exempt fund will be used solely for the benefit of those families whose members have been injured or killed by the COVID-19 vaccines including support of research through the Advanced Biological Research Group, ABRG.org.

# Foreword

## By Christiane Northrup, MD

I will never forget the first time I heard Dr. James Thorp speaking on the issue of Covid vaccines in pregnant women. It was during the heart of the Covid scam—soon after the experimental Covid shots were rolled out. What a revelation! I couldn't believe my ears as Jim revealed his findings. And, more importantly, that he had the courage to publicly share them whilst almost all other doctors were remaining silent.

Here was another doctor like myself, a Fellow of the American College of Obstetrics and Gynecology (ACOG), actually daring to share the devastating results he was seeing in his pregnant patients who had received the experimental Covid shots, encouraged to do so by ACOG. Even more astounding, Jim was willing to admit that he had been wrong in recommending the flu and DPT shots in the second trimester of pregnancy; a practice that started—much to my chagrin—in the 1990s.

Jim is a practicing expert in maternal-fetal medicine whose work has been widely published in academic medical journals. He has been above reproach for decades. What I was hearing from women via friends, family, and my social media platforms, Jim was seeing up close and personal in his busy practice—the devastating effect of the Covid shot on his pregnant patients. And once again, unlike the vast majority of our colleagues, Jim spoke up about it. Loudly! And brought out the data that so many others were trying to hide.

Tentatively, I reached out to Jim. Would he be willing, I wondered, to weigh in on a study that a group of us, led by Tiffany Parotto, started in the spring of 2021 tracking the menstrual effects of the Covid shot on those who were simply near recently inoculated people? Jim was all in. As a result, our group, which became *MyCycleStory.org*, has now published several papers on our findings. Bottom line: Indeed, the Covid shot transmits something to the unvaccinated in ways that we don't understand, but which has resulted in excessive bleeding, the passing of decidual casts, and even miscarriages.

Jim not only weighed in, he also provided masterful statistical guidance for our efforts. I half expected to be rejected by Jim when I first reached out given that I have been somewhat of an outlier in my chosen specialty long before Covid. (E.g., I came out against Gardasil on the Oprah show in 2006 when it was first rolled out—and have since seen teenagers become menopausal as a result). But Jim had a wide-open mind and a huge heart for doing the right thing. Not only did Jim "get it" about the Covid agenda, he introduced me to other Ob-Gyn's who were having similar experiences and daring to speak out. We became a very small, but very aligned group—realizing how badly we needed each other's support during a time of such deep deception and censorship from our colleagues, former friends, governments, and public health authorities.

As I was getting to know Jim, I discovered that he had long been influenced by the life of Ignaz Semmelweis, the Austrian doctor who dared to suggest that physicians should wash their hands between the time they did surgery on sick patients or worked on cadavers—and then—with blood-stained garments and instruments, went on to attend birthing women. At that time, women were dying in far greater numbers in hospitals—from so-called "child bed fever"—than those attended at home by midwives. The arrogant doctors of the time did not understand germ theory or sterile technique, and Semmelweis's suggestion that washing one's hands between procedures was ridiculous. Like so many who had questioned mainstream dogma, he was mocked and ridiculed by his colleagues, and committed to an asylum where he ended up dying. His theories were later proven to be correct—just as there is no doubt that those of us who question the safety of vaccines in pregnancy will be vindicated. This has already started.

But what of these women, whom Jim affectionately calls, "my patients"? What of the thousands of women who have lost their babies and/or their fertility as a result of this horrific Covid debacle? This is yet another example of how the failure to question authority within the medical/pharmaceutical-industrial complex costs women dearly. The specialty of Ob-Gyn has a long history of backing the wrong horse. For profit.

I was a newly minted Ob-Gyn when I joined a practice that specialized in helping DES daughters—women born to mothers who took the drug diethylstilbestrol in pregnancy to prevent miscarriage. Not only *didn't* it work, it resulted in infertility, abnormal pap smears, premature labor, cervical cancer,

# Foreword

and abnormalities of the genital tract in the offspring of the women who took it. Prior to that, thousands of children developed limb abnormalities or died prematurely because their mothers were prescribed a drug called thalidomide. In each of these cases, it took years before the adverse effects of these drugs were reported and the drugs were removed from the market. How does this keep happening?

The awful truth is that we are just waking up to the fact that Covid was yet another strategy designed to cripple the human race and depopulate the planet. Let me be very straightforward—I have come to the conclusion that the entire medical-industrial complex is, at its core, Satanic. It is run by a death cult. And yes, there are many innocent and wonderful people working in these fields who have no idea that this is going on. Fortunately, that is changing rapidly as we humans awaken to the truth.

Here it is: We are made in the image and likeness of God. When we align with God, we find the strength to stand up for what is Goodly and Godly. And as a result, we have the ability to create a new world—a world that represents Heaven on Earth. But this can't happen until we name the enemy and vanquish him. And that is where Dr. James Thorp and coauthor, pioneering journalist Celia Farber, come in. This book has been written "for such a time as this."

Standing up for life—for women—and for the future children of this planet. This volume you are holding in your hands is a Divinely inspired weapon, to say to Satan and all his minions: "Get thee behind me. Your rule on this planet has now come to an end." Amen.

# Introduction

*Some wise man once said that sin is that which is unnecessary.*
—*The Sacrifice*, Andrei Tarkovsky

*If the Bill of Rights contains no guarantee that a citizen shall be secure against lethal poisons distributed either by private individuals or by public officials, it is surely only because our forefathers, despite their considerable wisdom and foresight, could conceive of no such problem.*
—*Silent Spring,* Rachel Carson

*We never, ever give experimental treatments to pregnant women. We don't do that because 60 years ago, thalidomide taught everybody a lesson. Toxicity was built into these agents. . . . Five discreet mechanisms of toxicity . . . The conclusion is that we are under intense attack.*
—Mike Yeadon, former VP of Pfizer

## War on Life

This book contains parts of the unfathomable, dark history of how every regulatory health agency in the United States acted in complicity with a rogue transnational military government, to make "vaccines" against COVID-19 widely available without safety or efficacy testing. Pharmaceutical companies and governmental agencies colluded, unleashing a solid wall of government "money," including taxpayer money, in an unprecedented society-wide assault with the goal of mass injecting serums said to contain magical properties to "save lives." They maliciously maligned and banned simple, already approved, tried and true early treatments for "Covid," even forcing people to smuggle them in via laptops or sweater hems, to their dying loved ones in hospitals. They threatened pharmacists, they pulled medical licenses, they even got false medical "papers," produced by a front company, about hydroxychloroquine published in *The Lancet*.

They falsified not only medical journal publications, but also taxpayer-funded governmental databases.

Their "solid wall of money" soaked into all tiers of society: So-called "legacy media," TV, radio, print, social media, entertainment, rock music, even rap music—all pumped out relentless "Get Vaccinated" propaganda. Who knew that churches, mosques, synagogues, sports stars, celebrities, talk show hosts, and every imaginable "medical" or social structure in the US, starting in 2019, were absorbing these public health messaging funds?

The shocking numbers of this monetary colossus can all be openly found at www.USASpending.gov, branded by a depressing American flag logo, that appears to be shredded.[1]

By force of sheer cultural terrorism and propaganda, licensed by the ostensible "emergency," the FDA opened the floodgates to "Emergency Use Authorization" (EUA)—to "roll out" the most lethal product in medical history.

Reams, volumes, and books have already documented this grotesque, impossible to believe (though we lived through it) chapter of history.

Our focus, in this book, is on how this chemical assault program specifically targeted pregnant women and their unborn children.

My coauthor and I cannot tell such a story seamlessly or with consistent narrative structure, as it is too vast a crime to reverse engineer or narrate comprehensively across a smooth story arc. That may come decades from now. But, if anyone could achieve that kind of narrative cohesion, it would have to be those who were on the *inside* of this jet-black plot. One wonders if their souls are so indelibly surrendered to the cult of vaccinology that even *they* could not tell us, were they to fully confess, what it was all about. Vaccine cult agents seem to have adopted a near MK-Ultra level disassociation in the brains such that even mass deaths before their eyes do not translate into "vaccine hesitancy." It's reminiscent of Mao Tse-tung level mass mind control—cheering death as "progress" and part of the Great Leap Forward. America had fallen, at last, to a mass Illuminist trance.

This "medical catastrophe" (in quotes because that's only *one* of its aspects) has entirely different characteristics from previous ones, in a crucial way: This was not a "mistake"; it was planned. It was not the inevitable result of shoddy testing, greed, cutting corners—none of that. This was an altogether different animal: premeditated, malevolent, alien, AI–driven, and infused with revolutionary wrath. All familiar standards, values, and even known forms of corruption in medicine were thrown out, and replaced with this New Covid Doctrine, seemingly overnight.

Anyone who had any misgivings about the new Covid Vaccinology revolution was marked for destruction, their professional days numbered. Instead of doctors being able to lean on, cite, and argue data, stats or facts, they were confronted with a revolutionary cloud known as "misinformation." Anything other than believing that the "vaccine" was the most perfect of medical creations, anything short of blind worship, was deemed counter-revolutionary. Once a person—citizen, doctor, politician, or anyone—was branded an "anti-vaxxer" it was somehow encoded that they become marginalized, or worse. In some European countries, like Germany, for example, some doctors were sentenced to prison or, as in Canada, forcibly tortured and incarcerated in "insane asylums" for "spreading misinformation" about Covid. In the US, as well as across Europe, Australia, and New Zealand, doctors were stripped of their medical licenses, brought before courts, and in some cases, it appears, even killed for voicing opposition to the Covid mandates, lockdowns, mask mania, and especially, vaccine propaganda.

For example, Dr. Luke McLindon, formerly principal investigator at the Australian Institute for Restorative Fertility in South Brisbane, is a highly esteemed fertility specialist. He was very concerned about the Covid vaccines from the outset and kept exacting statistics for every single patient in his clinic, which specialized in helping couples who were struggling to get pregnant. His data was crystal clear: Vaccines were disastrous early in pregnancy. After the vaccines were rolled out, within just a matter of months he saw his early pregnancy miscarriage rate jump from 19 percent to a staggering 74 percent. As soon as he tried to publish this—his own data, from his own clinic—he was driven out of his practice.[2]

## The New World (Dis)Order

The Covid revolution, with all its infinite brutality, had been meticulously planned, for decades—not as a "medical" event but as an economic and population control event. That story is partly revealed through the comprehensive documentation of a US patent research expert, Dr. David Martin, who states in a *London Reel* interview:[3]

> Coronavirus, COVID-19, the whole thing was premeditated. It was murder. It was active terrorism by a state against the world. For fifty-eight years the

# Introduction

xix

> United States, the UK, in collaboration with researchers around the world planned to use coronavirus to instill the most tyrannical reform of society that this generation has ever seen.

For readers who are new to this, one can discern the planning and the "conspiracy" by researching the work of many historians and whistleblowers who have, since the 1950s, been trying to tell us: A global "cabal" is planning for a "New World Order" in which all the freedoms and values of the Old World will be systematically eliminated. There are to be no nations, no religions, no property, no privacy, no families, and no such thing as resisting the state's direct interjection, or penetration, of your skin—whatever they may elect to inject you with at a given time. We'll leave that subject to others, but suffice to say, central to all these historical narratives about the Club of Rome, the Committee of 300, the WEF, and so on, is that they want a *dramatic* reduction in the global population. One need not dig too deep to uncover these vast conspiracies against humanity.

The question, prior to 2020, was: What were they going to do to achieve this goal, this harrowing catastrophe?

And the answer was so surreal and dystopian, nobody imagined it: They were to launch a global terror campaign telling all people on earth that a new virus jumped species from bats to humans at a wet market in Wuhan, China—that this lethal new Coronavirus was spreading via every human interface. People must be "locked down" in their homes and exiled behind face masks, in addition to being asked to cease imagining coming within six feet of another human being, for any reason. And that included a dying parent in a nursing home.

The evidence is quite solid that far from Covid Vaccinology (the secular religion) being a bundle of "mistakes," the *intent to harm* was there from the beginning. Why else would they have made sure all the tried-and-true treatments that cured the syndrome of this extreme and unusual flu we came to call COVID-19 in 2020 were vilified, lied about, and banned? Doctors, nurses, and pharmacists even lost their careers for prescribing ivermectin or hydroxychloroquine, which had long-term, known safety records and were very effective in treating COVID-19. Instead, their goal was to herd terrified people into hospitals, separate them from their loved ones, cut off communication, and get them on ventilators (deadly, for Covid patients) and/or remdesivir (ditto).

Each of these stations of death had a steep bounty price attached to it; the "health care" provider, the hospital, and the entire system were soaked in the new money that came from the colossal $5 trillion Covid "budget." Financial incentives were offered for a positive Covid test (which were never able to identify any viral "infections"), administering a vaccine, admittance to the hospital, and more horrifically—for prescribing remdesivir and putting "Covid" patients on ventilators, both to disastrous effect.

How all this became possible is unanswerable—as evil remains essentially a mystery, throughout human history. But to target pregnant women and their unborn children is a rare level of evil.

It seems sadistic "innovations" in obstetrics that haunt women, never leave them alone. There have been many highly profitable "mistakes" and "disasters" in the dark history books of obstetrical care.

## Disaster #1: Diethylstilbestrol (DES)—The Worst Catastrophe in the History of Ob-Gyn, Before the Covid Vaccines Hit the Market

For at least thirty years, from about 1938 to 1971 and beyond, diethylstilbestrol (DES)—a synthetic form of estrogen—was prescribed to pregnant women to prevent miscarriage, premature labor, and related complications of pregnancy. The number of women taking DES is difficult to determine but is estimated to be more than 10 million. DES was a drug that crossed the placenta into fetal circulation and is now well established to be "an endocrine-disrupting chemical." The innumerable deaths and injuries that resulted is staggering. It could easily approach 50 million or more global citizens when one considers recurrent pregnancy losses, cancers, and other diseases DES has caused in multiple generations.

At the turn of the century, DES was classified as a carcinogen in humans—meaning it caused cancer. But that was only a minor part of the carnage. As it turned out, DES was a multigenerational curse. While the DES disaster primarily focused on women it also had severe effects on men. The drug affected not only the baby girls and baby boys that had been exposed to DES in their mother's womb, but also their own subsequent offspring for two or possibly more generations.

## Disaster #2: Thalidomide—Brought to Public Awareness, While DES Was Largely Swept Under the Rug

The "Titanic" of them all, before 2020—the one that seared itself into our collective trauma memory—was the thalidomide scandal of the late 1950s and early 1960s. That drug, developed in postwar Germany by the pharmaceutical company, Chemie Grünenthal—which was staffed with former

# Introduction

Nazis—was marketed as everything from a sedative, a sleep aid, and mood stabilizer, to a cure for morning sickness. At the peak of its popularity, it was advertised and freely sold not only in Germany, but in Canada and up to sixty European and African countries. It had been marketed all over the world with the ease of an air freshener, or frozen dinners, by the time it was discovered to have been the cause of severe birth defects, most horrifically, babies born with missing limbs.

The only country that did not approve this "wonder drug" was, surprisingly, the United States, where an astute Canadian pharmacologist and physician, Dr. Frances Oldham Kelsey, blocked it at the FDA for lack of safety data in pregnant women. Her very first month on the job, and her first assignment—it was given to her, ironically, with a little bit of dismissiveness. They had decided to start her on something "easy." Since over sixty countries had already green-lighted thalidomide, nobody expected the US to balk.

Dr. Kelsey saved the lives of hundreds of thousands of babies by her act of defiance against drug company and licensing company pressures. The dark impact of thalidomide was massive and became the very benchmark of the kind of medical disaster that would, ostensibly, *never* happen in today's world. In fact, thalidomide is credited with giving rise to what was supposed to be an enlightened age of *far stricter* FDA regulations. This is part of the thalidomide mythos.

Due to the worldwide dark publicity, impossible to deny, of the devastating photos of limbless thalidomide babies, the world experienced a crisis in drug testing and drug approval, along with the drug being taken quickly off the market when the effects were found out.

But thalidomide was a chemical, a crude thing nobody in their right mind would defend. The aftermath created the myth of a "never again" ethos: *Now* the painful lessons had been learned, and such a thing would never afflict pregnant women again. *Now* the "Golden Rule" would protect pregnant women in the future, and one presumed, would not be swept aside.

But in the Covid era, it was not a mere chemical being injected into pregnant women: the Covid shots were part of a dark revolution in "public health." Vaccines against Covid were not only deemed good, they were an essential part of the new public contract, of how to be a responsible American—or global citizen of just about any country, save North Korea. The Covid vaccine craze of 2021 had entirely eliminated the possibility that "vaccines" *could* be harmful. Frances Kelsey herself would have been powerless to prove otherwise were she still alive and at her FDA post in 2021. A "religion" had taken over with hundreds of billions of dollars attached to it, vaporizing the common sense and courage of all but the most defiant MDs. Those who spoke out against it, even with stark and damning evidence, were putting themselves in the line of fire: they lost jobs, careers, reputations, medical licenses, friends, colleagues, and even family bonds.

Opposing thalidomide was deemed valiant and heroic. Opposing Covid vaccines was labelled dangerous, filthy, and crazy. No bridge existed, by 2021, between the horrors of thalidomide and the horrors of the mRNA vaccines. The FDA even approved the Covid shots, unanimously, for infants as young as six months old. Not because they "worked," not because they were "safe," but because some inexplicable mass delusion had overtaken all the arms of the medical, public health, and safety testing professions.

Between 1962 and 2021, the pharmaceutical and biotech industries would grow into something unimaginable and unrecognizable. Parasitically fused with the government, the military, the media, and the educational system, by 2020, with a grand total of over $5 trillion of Covid spending allowed by Health and Human Services (HHS), the FDA was frankly nothing but a clearinghouse for anything Big Pharma and the Davos/Club of Rome globalists could dream up.

It was up to them who would die, how many, how fast, and how horribly. Never again would there be an admission, or compensation, or brakes on the freight train to hell. It would become a Pharma-dystopian dictatorship, and someone like Frances Kelsey, if she were to try something like that in today's world, would be either fired and maligned, threatened with de-licensing, or perhaps even killed.

By 2021, the Covid mRNA vaccine-gaslighting was so successful that the majority of Americans were thoroughly brainwashed into believing that there was *no* such a thing as a "dangerous drug or vaccine."

By 2021, not even nurses—even when a patient dropped dead before their eyes—would "admit" that, yes—the mRNA Covid shot was the cause.

This extreme ideological protectiveness of such a dangerous medical product can only be explained by an unlimited Health and Human Services propaganda budget, and a nationwide professional hostage crisis where each and every person working in our biggest single industry—health care—had been successfully gaslit into *not* believing their own eyes and ears.

That doesn't just require a "propaganda" budget, but decades of mind control and advanced psychological operations. We didn't even consider that they were working on our minds all these years, before Covid "hit," but they were. They knew us way better than we knew ourselves, and each other.

When Andrei Tarkovsky's character in *The Sacrifice* comments that "sin is that which is unnecessary," he quietly touches on the all-too-common human fallacy that *doing something* is always better than doing nothing. This is a quintessentially American idea. It is also the very bedrock of pharmaceutical profit: *Do Something*.

But, as this writing is focused primarily on the disastrous effects of the Covid "vaccine" on pregnant women, it is important to note that pregnancy is only safe when it is seen as a normal event. Covid propaganda rendered it a dangerous event, that required "heroic" interventions.

And the results, still ongoing, are ruinous.

# PART ONE

Chapter One

# Elixirs of Death: A Day in the Life of an Obstetrical Doctor

*You boast, "We have entered into a covenant with death, with the realm of the dead we have made an agreement. When an overwhelming scourge sweeps by, it cannot touch us, for we have made a lie our refuge and falsehood our hiding place."*
—Isaiah 28:15

*I am vengeful. I'm calling it the big kill. We've seen the biggest kill, ever, in medicine's history . . . How can you look into the eyes of a pregnant woman and tell her that this experimental product is safe? Any physician that has done that should be in jail.*
—Dr. Roger Hodkinson, interview with Robert Vaughan,
*Just Right Media*

## The Sonogram

My cell phone rang, and when I saw the incoming number, I did not want to answer. It was the cell phone number of one of my sonographers. A dread came over me.

"Dr. Thorp—"

She sounded very upset.

"Dr. Thorp, we have another fetal demise. I need you to talk to the family. Can we set up a Zoom call?"

The blood rushed from my head. Another one.

I sat down, and looked out my window at the bird feeder, where a starling had landed.

"Thank you," I said. "Let's go over the images now."

There was a silent detonation in my psyche. I briefly lost presence in my body, mind, soul, spirit, and voice. They were not connected.

I work as an Ob-Gyn and high-risk pregnancy subspecialist also known as a maternal-fetal medicine physician. In a typical day, I review between twenty and forty sonograms (ultrasound images of preborn babies or fetuses). In this case, it was my job to review the images and speak to the parents who were not—until that moment—my patients.

This was their first child—a boy. It had been a smooth, uneventful pregnancy. The baby was at about twenty-six weeks in her pregnancy, and there was plenty of amniotic fluid.

I reviewed the couple's vaccine history, and my heart sank. They had both received two Pfizer shots and one "booster." Their chart had them listed as "the best thing you could be: Fully vaccinated."

I felt absolutely helpless. This was the third baby that had died in our practice, close to birth, in a week. Why in the world is not everybody at this hospital acknowledging this calamity?

Sonographers are in a very tough position; they're at ground zero, you might say. When they see that a baby has died or any other abnormal finding, protocol dictates that they are not allowed to tell a patient the diagnosis—not tell the parents directly. Instead, they feign normalcy, make the mother comfortable, step outside the room and arrange for a physician to review the sonogram and deliver the news. That meant me.

Pfizer's own data pointed a straight arrow to this fact: it was the Covid shots that killed their baby.

I arranged the Zoom call and took a few deep breaths. This was not the time to tell them the raw truth—not even close. Honestly, they might become suicidal if they understood. Understood what? Understood what Pfizer's own data revealed about catastrophic outcomes in pregnancies for women exposed to their Covid "vaccines."

The only thing to do was to stay present with them as the crushing news sank in. To make matters even worse, the mother would have to go through a painful labor and delivery, knowing she would *not* greet a living baby when it was over. Childbirth is one of the most painful physiologic events in life, but the pain is assuaged by the mother's hope and elation when she experiences the baby's birth and first cry.

A preborn baby dying in-utero in the third trimester of a normal pregnancy is vanishingly rare. But now it was becoming a regular part of my life as an obstetrical specialist to interface with something so dark and sinister, to have to conceal it, even if it meant almost literally lying, when they asked the piercing question: "Dr. Thorp, why?"

Only during sleep could I escape this suddenly darkened, malevolent post-Covid world, where several babies a week were turning up dead in the third trimester of an otherwise perfectly normal pregnancy.

I can't quite describe to you how aberrant this all is. Inconceivable, is actually the word.

I always see grieving parents face to face when we lose a baby—in person or, if they're in another state, as was the case this time, on Zoom. It's always the hardest part of my job.

Once we were connected, I introduced myself briefly.

I needed to say it right away so as not to drag it out and worsen their trauma and shock.

"I'm very sorry but your baby has passed in the womb and there is no heartbeat."

I said nothing more, but maintained eye contact, and prayed silently. I remember that she placed both hands over her mouth.

When parents are in this devastating situation, the moment they get the news, I've seen a range of immediate reactions. It can be shocked silence, loud wailing—it can also be blind fury. In the case of this couple, they just stared at me in total silence. She began weeping and through her tears she finally said a few words that are known in grief literature as "bargaining."

"This is not possible, Dr. Thorp. The baby was moving yesterday. He was kicking, really strong kicks."

The husband said nothing. He just sat staring at the screen. He seemed motionless, as though afraid to move, frozen in time.

One thing I will never know, is how it feels to carry a life; never mind what courage it takes to do what that young woman did that day; to sit on a Zoom call and hear this. I prayed silently for mercy, grace, and solace. I could feel this couple's horror and pain in my own body. They had been two happy people expecting the most joyous event of their lives to unfold, and now they were broken, shattered in pieces—utter, complete brokenness.

The room behind them seemed to have been overtaken by gloom, a spiritual blackness, like the vestibule of an abandoned haunted house.

The husband spoke for the first time. "How did this happen?" He asked, "What could possibly have gone wrong? We've been so careful."

"We don't know," I said, "but I can assure you it was nothing you did."

They were moving between denial, bargaining, anger, depression—all the known stages of grief. My words were all they had to hold on to, and I had to choose them very carefully.

"Dr. Thorp, are you sure?" the father further inquired.

He looked at me with pleading eyes.

"I'm sure," I said. "I'm so sorry."

And then he asked, "When did our baby die?"

I gently shared that I thought the baby was alive yesterday, as evidenced by the fetal movement and by the ultrasound findings of the completely normal amniotic fluid, and no "postmortem findings" on the ultrasound images of their stillborn in-utero.

Later that day, I would call their primary obstetrician to deliver the news, knowing full well this would have been the doctor who failed to warn them about side-effects of the "vaccine" she had prescribed.

In these solemn situations, I try to gain insight into the parents' family, friends, and social support networks, and ask if they have a faith tradition. This couple shared that they were Christians and yes, they wanted me to pray for them.

I gave them my cell phone number and encouraged them to call me any time, if it would help them to talk or ask questions in the coming days and weeks. My secret hope was that in a few months, they would call back and ask: "Dr. Thorp, could this have been caused by the Covid shots?"

If they accepted that devastating blow, they could at least be protected from taking any more shots in the future. On the other hand, their regret and guilt could be incapacitating. If I had more time, I believe, I would be able to let them know this was not their fault. Expecting parents were being assailed with a propaganda operation so carefully tailored, based on decades of advanced study of human nature. It was like a predictive behavioral algorithm. The base of the programming—the first piece of code—was a work of dark brilliance:

"My Ob-Gyn would never recommend this if it could harm my baby."

That belief, it turned out, was the first and fatal mistake. What they didn't know, was that sixty thousand Ob-Gyns had been bought and paid for by way of Health and Human Services and Centers for Disease Control and Prevention (CDC) to The American College of Obstetricians and Gynecologists (ACOG).

They never did call me with that question. Most do not.

I've been in clinical practice of obstetrics as a physician since I graduated from Wayne State University School of Medicine in 1979.

Obstetrics is known as the "well patient practice." Put more simply, it's the field of medicine that brings the most joy. We were the "lucky" doctors, who rarely, unlike our colleagues in other fields, had to come face to face with death. It happened, of course, but very rare.

Fetal death is defined as a preborn baby (fetus) who dies in the womb at or after twenty weeks. Compared to miscarriage, fetal death is much rarer and with the progressive improvements in obstetrical care and fetal surveillance it has steadily declined in the US over the past century. By the US national statistics, the fetal death (stillbirth) rate was about 5.8 deaths for every 1000 births, prior to Covid.

In fact, the first year of Covid, 2020, the fetal death rate actually dropped. In the year of "raging Covid," the stillbirth rate dropped to 5.74 per 1,000 births according to US vital statistics (Statista). So "COVID-19" never increased the risk of fetal death/stillbirth by itself.

But, after the Covid "vaccine" was rolled out, post-partum nurse and whistleblower, Michelle Spencer, reported that her hospital sent all nurses in her facility an email documenting twenty-two-plus still births per month. By Spencer's account, they had experienced only one to two stillbirths per quarter, before that. Many other areas of the world experienced similar rates, especially in the more vaccinated locations. At least three locations in Canada, two in Ontario (Peterborough and Waterloo), and one in British Columbia (Vancouver) documented unparalleled rates. In the Vancouver facility, five independent whistleblowers documented thirteen stillbirths in just one twenty-four-hour period, attested to by Dr. Mel Bruchet and Dr. Daniel Nagase, in addition to three independent doulas.

These surges in fetal death after the vaccine rollout represent a forty-plus sigma event—an unprecedented rate. A sigma event (standard deviation) is a statistical term of variability; in this example, assumed to be about 0.5 fetal deaths per 1,000. Financial analyst and Covid vaccine whistleblower Ed Dowd calls this a "Black Swan event." In air travel, this would be the equivalent of a jetliner crashing two or three times a week, while the airline industry pretends nothing unusual is happening.

## Where Was Your God?

*For you created my inmost being; you knit me together in my mother's womb.*

—Psalm 139: 13–14

Non-believers will ask: "And where was your God?"

This is my answer.

He was there when the baby was made, fearfully and wonderfully, and when it grew—the beginning of hands and fingers at eight weeks, teeth beginning to form at week nine, lips and ears by week sixteen, and the beginnings of soft-fuzz hair at eighteen weeks. At twenty-seven weeks, preborn babies open their eyes, and blink.

God created that baby, perfectly.

His enemy, however, came to kill, steal, and destroy his life. God does not stop or prevent evil; he only gives us life along with free will. The enemy, however, sets endless traps to take us away from God's will; and the Covid crimes may be the enemy's single, greatest trap yet—his final rampage.

I know the rhythm of life. I've developed a feeling for it.

Each pregnancy is like a sea voyage—the embryo as precious cargo that grows so fast, stronger and stronger, until, after twenty weeks, we can let our guard down; we can see the shoreline. Something (life's plan itself) is always protecting the baby, rocking it gently with every heartbeat closer and closer to the shoreline of birth. It works. It holds.

When something goes wrong, we can usually intervene, we have space, a berth—a dialogue with life, with creation, and God's will. He created the preborn in the womb to be durable, to endure amazing capacities to hold on to life, even in very precarious situations. The COVID-19 "vaccine" was not one that countless embryos and fetuses could withstand.

I've delivered thousands of babies over the decades. I've transfused many preborns in the womb that were at death's door with severe anemia. I've watched God's miracles slowly bring them back to life—answering my prayers and guiding my hand while placing a very long needle into a tiny fetal blood vessel; sometimes more than six inches away from the mother's skin, using ultrasound guidance. These amazingly resilient preborns are able to withstand this extreme procedure along with blood transfusions, administered very slowly and carefully.

I know how much "give" they have—these miracles of life. Many of these preborns who triumphed over these procedures continue to communicate with me to this day.

Most of the time, between the Creator and my own set of skills and instincts, I can save a baby. But after 2021, I couldn't. I was absolutely helpless and without tools. Everything had changed, and what I am calling the "give" was not there anymore. It had been displaced by "a presence" of something alien, metallic, synthetic, causing utter chaos in the whole blueprint of life.

The enemy had advanced his murderous agenda and was now reaping souls in the womb. Souls who were wanted, growing perfectly, with no genetic or other challenges. Why were they dying?

This brings us to the beginning, the alpha and the omega, the number one principle of obstetrical care known as the Golden Rule. They used to drill this into us in medical school, and there was never any question about it.

The Golden Rule is simple: Never expose a pregnant woman to any untested substance for any reason.

The word never, meant *never*.

## Chapter Two

# Nurse Bearing Witness: How a Delivery Ward Became a Death Ward in 2021

*And the babies continue to die. I interviewed Dr. James Thorp . . . I interviewed two independent midwives. And all confirmed that they are seeing unprecedented horrors in the birthing rooms: compromised placentas with networks of calcifications . . . placentas that are shrunken, or flat, so that the babies cannot grow normally and must be delivered prematurely; babies with congenital malformations. Babies with breathing problems—air sacs between their chest walls and their lungs, which condition was identified in the Pfizer documents. Deaths in childbirth are up 40 percent, as women are endangered by amniotic sacs that fall apart and cause bleeding. Women have been returned to the unsafe childbirth conditions of the nineteenth century.*
  —Naomi Wolf, *The Pfizer Documents* and *Facing the Beast: Courage Faith and Resistance in a New Dark Age*

As a specialist in pregnancy complications, I no longer deliver babies. The ones who were truly witnessing the Covid carnage with their own eyes and hearts every single day were the labor, delivery, and postpartum nurses; the ones on the wards whose job was taking care of mothers, before, during, and after birth.

My mother Mollie was a labor and delivery nurse. They're the ones who really know what goes on, what happens to women in childbirth, what normal, healthy births look like, and certainly what an extreme "danger signal" looks like. The kind of "signal" that so tormented Ignaz Semmelweis at Vienna General Hospital in the 1840s. At that time, mothers were dying of a dreadful infection in childbirth at alarming rates. Ignaz Semmelweis is one of my heroes, and you'll learn more about him in a later chapter, but, in 2021 and 2022, it was primarily babies dying. Mothers died in childbirth also, but the babies were dying as though under a Biblical curse. It was pre-modern—something out of line with the direction of history. Yet, it was happening here, in clean, sterile, well-lit, American hospitals.

Starting in the year 2021, *they*—with Dr. Rochelle Walensky, director of the CDC at the helm—pushed, through every available mass media, and through the entire network of Ob-Gyns, the untested, lethal Covid shots for pregnant women. Walensky must have known that the COVID-19 "vaccines" were the deadliest and most injurious drug ever rolled out to the public. Everyone who did their due diligence knew—it was leaked by a whistleblower to the entire world in early 2021.

## Front-Line Witness to the Carnage

Michelle Spencer, a postpartum registered nurse at a hospital in Fresno, California, had her eyes wide open about the deadly lies of Covid, after her own mother was killed by a lethal hospital Covid protocol. Spencer's mother was her best friend and closest ally. When her mom got sick, from something unrelated to Covid, she became an unwitting victim of the murder-by-hospital protocols that seemed to be lying-in-wait like sleeper cells across the country, starting in the late winter of 2020.

"They treated her like she had Covid," Spencer says, in a June 2024 interview with me and Celia. "Even though I'm an RN, and I was treating Covid at the time, they wouldn't let me in [the room] to be with her. No matter what I said, they would not relent."

Michelle's mother was subjected to the weakening, eventually deadly protocols, including remdesivir, from which she finally succumbed.

"They murdered her," says Spencer. "That's what radicalized me. My mom and I were so close. She was my best friend. Before she died, she told me, over and over, whatever I do, *not* to quit my job as a labor and delivery nurse.

"'Those babies need you,' she pleaded. 'You cannot quit.'"

Fresh in her grief, Michelle's nightmare had only just begun. She noticed the effects of the deadly Covid shots in the pregnant women who were her patients, and their babies, and—something she had never before witnessed—a dramatic increase in babies born dead on the ward.

And, when Michelle would check the mothers' charts, without exception, they had recently had one or more Covid injections. On a Zoom panel discussion, Michelle recalled when she began to see the heartbreaking pattern.

> Before the shots, we saw no death or destruction—nothing. The protocol was to test them all on the PCR test for Covid, . . . some tested positive, but we saw no Covid symptomatically, and certainly no deaths. It all began after the shots were introduced. So, to give you some context, before 2021, we had maybe four fetal demises per year. After the shots were rolled out, by March of 2021, there was an *avalanche* of destruction. Every time I'd come to work, I'd see a dead baby . . . on [our] floor. Every day, another one, a soon as I clocked in and looked at the charts: Fetal demise. That means the baby dies anywhere between twenty weeks and full term.
>
> Around this time, we also went from having fifty babies in the NICU (Neonatal Intensive Care Unit), to having, by March of 2021, around eighty. It's a very large NICU.
>
> We also started to see mothers coming in to deliver their babies full term and the baby had already died inside. All of this is happening at once, in March of 2021. It was crystal clear. They would receive the Covid shot, and a couple of days later the baby would die. It was horrific.
>
> I would speak up about it and people around me just . . . said it's pesticides, someone else said it's something in the water. People didn't want to say it's the vaccine . . .
>
> I kept coming to work because I'm the breadwinner of the family. I had no choice really. I'm going to work and every day I'm still seeing these babies dying, all the time. But I kept hearing my mom's voice in my head saying, *don't quit*. Those babies need you. Those moms need you . . . you're all they have.

## Nurse Turned Whistleblower

On September 8, 2022, Michelle received a staff email that changed the course of her life. After reading the email that morning, her ensuing shock and rage transformed her from a deeply concerned nurse, into a front-line fighter and whistleblower.

While on the one hand, the email admitted, in plain text, that a very high number of babies were dying, at the very same time, it obliterated the normal context of responding and reacting to such a thing. It obliterated the natural four-alarm level of emergency that should have been present, addressing potential causes of the suddenly spiking deaths. It also took for granted, in familiar tones of Covid-era psychopathy, that no one was even upset about the deaths; rather, some things needed to be straightened out about the protocol for handling dead babies.

The email is reproduced here:

*Thursday, September 8, 2022 7:34 PM*
*Subject: demise handling*
*Good evening everyone,*

*Well, it seems as though the increase of demise patients that we are seeing is going to continue. There were 22 demises in August, which ties the record number of demises in July 2021, and so far in September there have been 7 and it's only the 8th day of the month. Now these statistics include [redacted], so you haven't seen all of them, and some have gone through the [redacted], but there still have been so many in our department. It's a lot of work for you as the bedside [redacted] and it's also a lot of work for me. Demises have taken a lot of my time away from the other groups of patients that I serve, so I hope this trend doesn't continue indefinitely. I know of a few more that are scheduled to deliver in the week ahead, so unfortunately, the process is going to be very familiar with all of you. Once again, I do so appreciate the time and attention that you give to the patients. When I follow up with them, they remember your names and the way you helped them get through a very difficult time.*

*We have recently had a few, less than 20-week demises [fetuses younger than 20 weeks] whose parents requested an autopsy. They can request an autopsy on these babies, however the baby still goes to [redacted] examines every baby less than 20 weeks born without signs of life, but it is only an external exam. For an internal exam, which is what the autopsy is, you will need to have the parents sign an autopsy consent, so send it along with the baby to [redacted].*

*To make this long story shorter, please follow the procedure in the fetal demise binder and do not let other departments tell you how to handle the specimens. The [redacted] involved has been doing the right thing, but was told by several different people to just put the baby in a body bag, so she did. There are a couple of things that I want to reinforce.*

1. Babies that are going to pathology are always small enough to go in the large white buckets. I know that it feels disrespectful to many of you to pour a bottle of saline over the baby, so you can wrap the baby in a saline soaked chux if it feels better to you, but it must go in a bucket if it goes to pathology.
2. Small babies going to the morgue can also be placed in a large white bucket with saline or a saline-soaked chux.
3. [redacted] informed me that they are no longer allowed to carry specimens in large paper bags, so place the placenta (or large white) bucket in the large biohazard bag only. Why they are not allowed to transport things in a paper bag, I did not ask, but that is what I was told.

*Thank you all so much!*
*Fresno, CA [redacted]*
*Phone: [redacted]*
*Fax: [redacted]*
*Mobile: [redacted]*

Recounting her experience, Michelle said,

When I read it, I thought, *I cannot believe* what I'm reading right now. It's all right there, for anybody wondering, "Did babies really die from these shots?" It says, in the first line: . . . in the month of August of 2022, there were twenty-two fetal demises. And that number ties with the record number of fetal demises in July of 2021.

So, if you're saying there's twenty-two dead babies this month and then there's also around twenty dead babies in July of the last year. And you take that whole year . . . That's, I don't know . . . 240 dead babies over the last year?

To give a context, she shared again that the average number of babies dying prior to 2021 was about four per year.

That's *insane*. It's a smoking gun.

But then, what was so incredibly surreal about it, is that rather than address *why* this was happening, [the email] was focused on what was referred to as "policies." Because of the increase in fetal demises, "we need to brush up on our policies" and learn how to handle a dead baby's body.

That was the word they always hid behind: "policies." I'm *so disgusted* by the word "policy," hearing people say, "Oh, you need to follow the policy. That's all I heard when my mom was dying. It's too bad you're a registered nurse and you already take care of Covid patients. We can't let you in. It's our *policy*.

She continued.

> . . . so, I had a reaction. No! I'm not brushing up on any policies—you guys need to figure out what the hell is going on.
>
> I took that email and sent it to a friend who is affiliated with Children's Health Defense. And from there, [we] met up with Matt from *The Epoch Times,* who connected me to Dr. Thorp. Thank God, because I believe that email was like a cry [from God] for help.
>
> I was [interviewed] on *The Highwire* with Del Bigtree, and CHD TV, and the story broke in *The Epoch Times* and elsewhere. It was as if I was the only normal human left, and yet I appeared [to my colleagues] like a troublemaker, for reacting.
>
> So many nurses were in a trance at this time, and you could really feel, sense, and see the trance. They didn't read that email and say, "Oh, my gosh—this email. These babies are dead *from* these vaccines."
>
> No one else did anything with that email.

Why wouldn't the administrators at Michelle's hospital investigate this catastrophic increase in stillbirths observed in their own facility? One must assume that this appalling callousness is tied to the fear that any on-the-record action taken that could connect vaccinated pregnant mothers to a stratospheric rise in stillbirths, could dry up federal and state funds to the hospital and/or be a career-killer for anyone who initiated the paperwork.

"I don't understand how people just brush it off and pretend it's not a big deal," Spencer said during our discussion. "It's absolutely a big deal. I do truly believe that babies are given to us by God, they're meant to bring us love and joy, and if something comes between that, trying to take your baby away, to me that's absolutely evil."

"I really think a lot of people are in a trance and they just don't want to admit that the vaccine could possibly be harming all these moms and babies," Spencer said in an interview with *LifesiteNews*. "A lot of them, they don't see it . . . because so many of them are programmed to think that vaccines are good when they're not."[1]

The ghoulish, full-length email has been published on several platforms.

In another interview we did together for CHD TV, Michelle again spoke to the eerie silence, denial, and internal censorship that had stopped her colleagues from uttering anything about this atrocity, even to each other. Maybe they did not even admit it to themselves; perhaps it was an internal protective mechanism from the severe cognitive dissonance they must have adopted in order to keep functioning. Or, perhaps many recognized the horror as well, but did not have the ethical, moral, and intellectual courage that Michelle possessed, and feared for their job security.

To this point, Michelle has documented on multiple platforms publicly that her institution was in fear of firing her. What did they fear? Likely their fear was very real and emanated from their own legal counsel. They knew multiple organizations backing Michelle possessed "deep legal pockets." If Michelle sued them for retaliation of her whistleblowing, the hospital would draw further national attention to their "killing fields" and their quid-pro-quo arrangement of monies received from governmental organizations. The gravy train could dry up. Also, a lawsuit could open another "can of worms" for the hospital as the legal discovery process would be catastrophic for the hospital and government.

## An Investigation Ensues

When Michelle's employers found out that she had been featured in the *Died Suddenly* documentary, which was released November 2022, they initiated an investigation. On December 4, to be exact.

"I thought I was going to get fired that day," she said in our more recent interview. "I prayed before I walked into that meeting. 'God, I know you're with me. You've been with me this whole time. Whatever is meant to be will be, but *be* with me in that meeting' . . . I was ready to give them my badge and everything."

Sharing her story of the meeting, Michelle continued,

> I was aware they were recording me. At the beginning, they said: "Are you aware that it's against company policy to copy and paste an internal document?" I responded by saying, "Why are you investigating why I copy and pasted an email? You guys need to be investigating why our babies are dying."
>
> After I asked that question, they sort of looked at me and said: "We're not here to answer your questions, Michelle, but you can go back to work now."
>
> I thought, "Wait, so I'm not getting fired? I was blown away."

A few weeks later, however, the hospital did issue Michelle's punishment—in the form of money withheld.

That year, when it was time for the annual retention bonus that all nurses received, Michelle was singularly excluded. When she brought this to the attention of the hospital administration, they stated that her personal retention bonus check was withheld because she had "broken the rules" by sharing an internal email with the public. "Well, that seems a lot like retaliation to me," she countered.

But here is a disturbing twist to the amount of the bonus paycheck that all the other "silent nurses" received. A stark "coincidence." Michelle verified this to me—not only through one nurse colleague, but multiple. She also shared it on multiple public platforms. It's quite stunning.

The amount of the annual retention check for the other nurses was $6,666.00. I'm sure many readers will grasp the symbolism of that number. Einstein stated that coincidence is God's way of staying anonymous. I believe Einstein is correct.

My response? I immediately set up a donation account for Michelle, from my X (formerly Twitter) account @jathorpMFM. My goal for this fund was to provide Michelle with a bonus check of $7,777.00 and this was successfully achieved. There is no such thing as coincidence. Love overcomes hate; Good overcomes evil; God always prevails.

I'm suspecting that many readers will be wondering how this could happen, as it's hard to conceive of evil occurring at this scale in our modern-day world. It happened because they allowed it to happen.

## What Pfizer Knew—Pfizer's 5.3.6 Legally Mandated Post-Marketing Adverse Event Analysis

> *In her Second Amended Complaint (SAC), Ms. Jackson details her knowledge of the particular circumstances in which Pfizer engaged in clinical trial fraud to induce FDA's issuance of the EUA. Among other things, this includes a trial design to avoid disclosure on immunity and transmission; short cutting the study to conceal negative efficacy and serious adverse events; manipulation of inclusion and exclusion determinations to reach predetermined results; unblinding of subject status and then falsely reporting the occurrence or non-occurrence of adverse events based on the subject's disclosed status; and suppression of available alternatives (Ivermectin).*
> —Pharmaceutical whistle-blower, Brook Jackson—excerpt from her lawsuit,[2] via Twitter/X

The FDA's post-marketing, safety data collection requires companies to share their oversight—their drugs' adverse events—and make them available to the American public. Pfizer's ninety-day post-market analysis began on December 1, 2020, with the shipment of the COVID-19 shots—and continued until February 28, 2021. As a practical matter, the actual injections were not started until ten to fourteen days after shipping; the actual Pfizer 5.3.6 was not twelve weeks, but only about ten weeks (seventy-four days). Pfizer's post-market analysis was completed on February 28, 2021. It included 42,086 "adverse events," including 1,223 deaths in about ten weeks. That makes this shot the *most lethal* and the *most injurious* medical product ever rolled out to the public in the history of medicine.

## The Attempt to Hide Incriminating Data

HHS/CDC/FDA colluded with Pfizer to bury the incriminating Pfizer 5.3.6 post-market analysis for seventy-five years. No disclosure for seventy-five years. That, in and of itself, is highly suspect. Fortunately, a whistleblower from either Pfizer or HHS/CDC/FDA released this formal report in early 2021, and it became available to everyone in the world who was motivated to do their own independent research and due diligence. Unfortunately, few did.

Later, a landmark Freedom of Information Act (FOIA) request, made by attorney Aaron Siri, yielded a judicial order mandating the release, and ironically, this occurred April Fool's Day (April 1) 2022. It was an identical document released by the whistleblower over a year earlier.

Based upon two "bates stamps" on each page of this document and UK spellings, I suspect that this document originated from the major Pfizer research facility in Sandwich, UK. In typical pharmaceutical tactics, the data is presented in a disjointed fashion, misleading, and poorly written, whether by intent or by ignorance. The writers lacked specific knowledge of obstetrical terminology.

Page 12 of this document, depicted below, describes 274 preborns in 270 pregnant mothers (some with twins) who were given the Pfizer's COVID-19 "vaccine," even though pregnant women were not allowed in the earlier studies. The obstetrical outcomes were an unmitigated disaster.

First, they chose to exclude outcomes of 238 of the 270 total outcomes. Why would they exclude most pregnant women from this report? Pfizer cited 124/270 (46 percent) of pregnant patients that experienced adverse events from the shot. Of the thirty-two pregnant women's outcomes that Pfizer chose to report, there was an 81 percent (26/32) miscarriage

Nurse Bearing Witness 19

## Pfizer 5.3.6 Post-Marketing Data (Page 7)

The deadliest and most injurious vaccine / medicine / drug **EVER** rolled out in the history of medicine

In Just 10 weeks (Dec 14, 2020, to Feb 28, 2021) there were 42,086 casualties including 1,223 deaths

Female-to-male odds ratio 3.26 with a 95% confidence interval 3.21 - 3.30

On X @jathorpmfm and DrJAThorp.com

BNT162b2
5.3.6 Cumulative Analysis of Post-authorization Adverse Event Reports

Table 1 below presents the main characteristics of the overall cases.

Table 1. General Overview: Selected Characteristics of All Cases Received During the Reporting Interval

| Characteristics | | Relevant cases (N=42086) |
|---|---|---|
| Gender: | Female | 29914 |
| | Male | 9182 |
| | No Data | 2990 |
| Age range (years): | ≤17 | 175[a] |
| 0.01 - 107 years | 18-30 | 4953 |
| Mean = 50.9 years | 31-50 | 13886 |
| n = 34952 | 51-64 | 7884 |
| | 65-74 | 3098 |
| | ≥75 | 5214 |
| | Unknown | 6876 |
| Case outcome: | Recovered/Recovering | 19582 |
| | Recovered with sequelae | 520 |
| | Not recovered at the time of report | 11361 |
| | Fatal | 1223 |
| | Unknown | 9400 |

# Pfizer 5.3.6 Post-Marketing Pregnancy Data Page 12

- 270 pregnant mothers
- 238/270 (88%) had NO follow-up
- 124/270 (46%) with complications
- 25/32 miscarriage (spontaneous abortion)
- 1/32 miscarriage (missed abortion)
- 26/32 (81%) total miscarriage rate
- 1/32 fetal death (stillbirth) rate of 31/1000 with an expected rate of 5.8/1000 – 5X
- 1/32 neonatal death rate of 31/1000 with an expected rate of 3.9/1000 – 7.9 X
- Breastfeeding complications in 17/133 or 12.8% of babies

*The deadliest, most injurious drug ever rolled out in medicine was pushed in pregnancy*

James A Thorp MD on X @jathorpmfm and DrJAThorp.com

---

Pregnancy cases: 274 cases including:

- 270 mother cases and 4 foetus/baby cases representing 270 unique pregnancies (the 4 foetus/baby cases were linked to 3 mother cases; 1 mother case involved twins).
- Pregnancy outcomes for the 270 pregnancies were reported as spontaneous abortion (23), outcome pending (5), premature birth with neonatal death, spontaneous abortion with intrauterine death (2 each), premature birth with neonatal death, and normal outcome (1 each). No outcome was provided for 238 pregnancies (note that 2 different outcomes were reported for each twin, and both were counted).
- 146 non-serious mother cases reported exposure to vaccine in utero without the occurrence of any clinical event. The exposure PTs coded to the PTs Maternal exposure during pregnancy (111), Exposure during pregnancy (29) and Maternal exposure timing unspecified (6). Trimester of exposure was reported in 21 of these cases: 1st trimester (15 cases), 2nd trimester (7), and 3rd trimester (2).
- 124 mother cases, 49 non-serious and 75 serious, reported clinical events, which occurred in the vaccinated mothers. Pregnancy related events reported in these cases coded to the PTs Abortion spontaneous (25), Uterine contraction during pregnancy, Premature rupture of membranes, Abortion, Abortion missed, and Foetal death (1 each). Other clinical events which occurred in more than 5 cases coded to the PTs Headache (33), Vaccination site pain (24), Pain in extremity and Fatigue (22 each), Myalgia and Pyrexia (16 each), Chills (13) Nausea (12), Pain (11), Arthralgia (9), Lymphadenopathy and Drug ineffective (7 each), Chest pain, Dizziness and Asthenia (6 each), Malaise and COVID-19 (5 each). Trimester of exposure was reported in 22 of these cases: 1st trimester (19 cases), 2nd trimester (1 case), 3rd trimester (2 cases).
- 4 serious foetus/baby cases reported the PTs Exposure during pregnancy, Foetal growth restriction, Maternal exposure during pregnancy, Premature baby (2 each), and Death neonatal (1). Trimester of exposure was reported for 2 cases (twins) as occurring during the 1st trimester.

Breast feeding baby cases: 133, of which:

- 116 cases reported exposure to vaccine during breastfeeding (PT Exposure via breast milk) without the occurrence of any clinical adverse events;
- 17 cases, 3 serious and 14 non-serious, reported the following clinical events that occurred in the infant/child exposed to vaccine via breastfeeding: Pyrexia (5), Rash (4), Infant irritability (3), Infantile vomiting, Diarrhoea, Insomnia, and Illness (2 each), Poor feeding infant, Lethargy, Abdominal discomfort, Vomiting, Allergy to vaccine, Increased appetite, Anxiety, Crying, Poor quality sleep, Eructation, Agitation, Pain and Urticaria (1 each).

rate, a rate that rivals that of the abortion pill RU-486 (mifepristone). Mifepristone carries a **BLACK BOX WARNING** from the FDA—their most dire danger warning to the public. Pfizer reported a 1/32 stillbirth rate (fetal death at or after twenty weeks), a rate that equates to 31/1000 birth that is five-fold greater than that expected from US national data from Statista of 5.8/1000. Pfizer reported a 1/32 neonatal (newborn) death rate of 31/1000 births—a 7.9-fold greater incidence than expected from US national statistics of 3.9/1000 births. Pfizer reported a 14.7 percent (17/116) incidence of breastfeeding complications in infants born of mothers who had the shot in pregnancy.

Pfizer's own data confirms that the deadliest, most injurious drug ever rolled out in the history of medicine was pushed to the most vulnerable: pregnant women, preborns, and newborns. I wished I had been wrong in 2020 when I repeatedly stated that if used in pregnancy the COVID-19 "vaccines" would make thalidomide or DES look like prenatal vitamins.

## It Gets Worse

The deadliest drug ever rolled out to the public is bad enough, but arguably there could be worse news. Astoundingly, the COVID-19 "vaccines" are, by far, the most *injurious* drug ever released on humanity. The "injure-to-kill" ratio is easily calculated from Pfizer's own report pictured below. Pfizer's "injure-to-kill" ratio is unprecedented at an astonishing 33.4 (42,086–1,223/1,223).

This means that for every person killed by the vaccine, 33.4 are injured. In historical medical context, the deadliest drug *ever* rolled out before the COVID-19 shot was thalidomide (discussed in a later chapter) that killed or injured 100 percent of preborns if given at the critical time in pregnancy. About 20,000 were injured and about 80,000 killed, yielding an injure-to-kill ratio of 0.25.

In historical military context, the two atomic bombs dropped on Hiroshima and Nagasaki in August 1945 injured about 94,000 and killed about 105,000, yielding an injure-to-kill ratio of about 0.9. Every major military conflict since has had an injure-to-kill ratio in the range of 0.9 to 5 with one outlier around 7. Was it by accident that over 13 billion shots of this "vaccine" were administered to over 5.3 billion people globally? Was it by accident that this "vaccine" is not only the deadliest, but also the most injurious with an unprecedented injure-to-kill ratio of 33.4?

## Pfizer 5.3.6 Post-Marketing Data Page 7

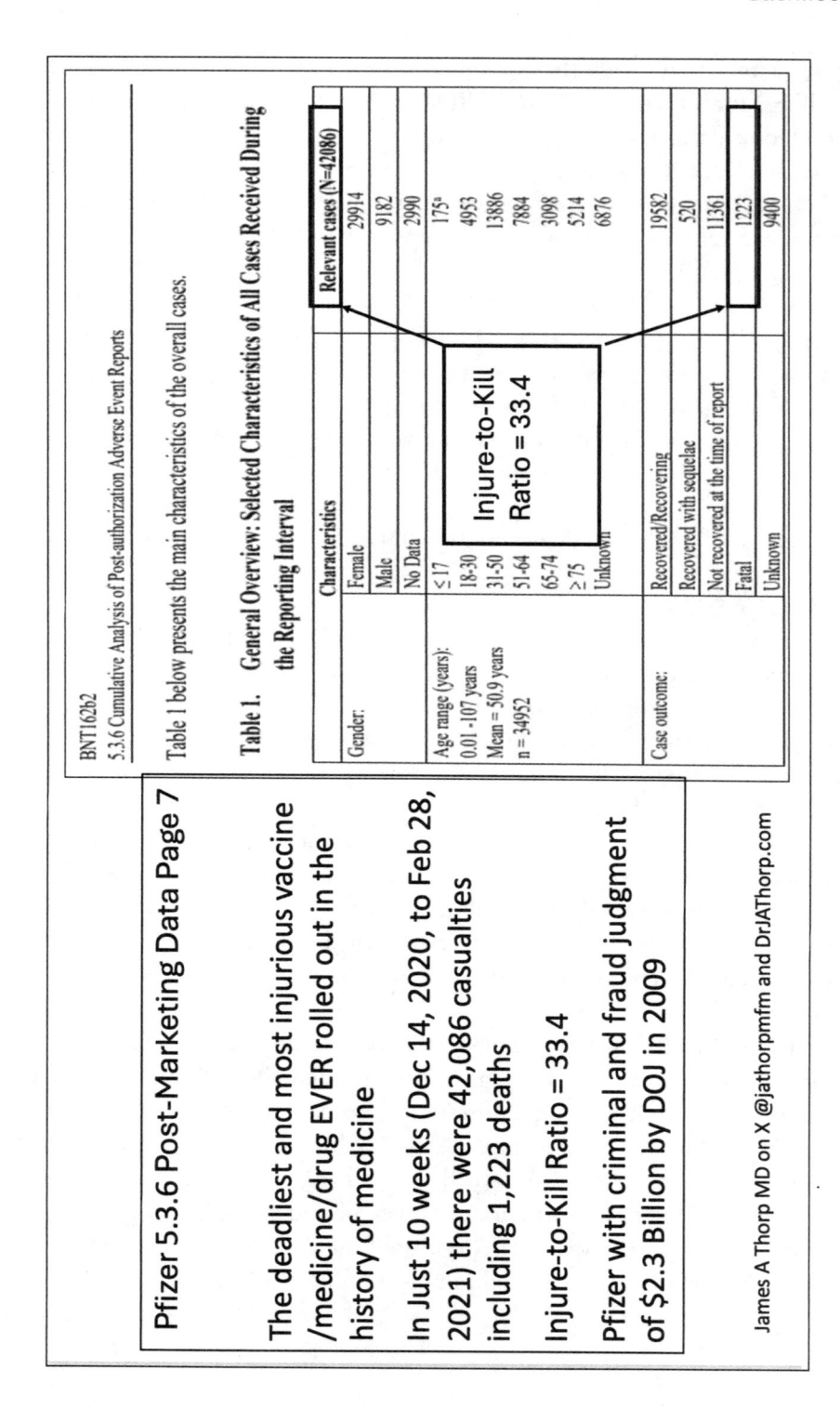

The deadliest and most injurious vaccine/medicine/drug EVER rolled out in the history of medicine

In Just 10 weeks (Dec 14, 2020, to Feb 28, 2021) there were 42,086 casualties including 1,223 deaths

Injure-to-Kill Ratio = 33.4

Pfizer with criminal and fraud judgment of $2.3 Billion by DOJ in 2009

James A Thorp MD on X @jathorpmfm and DrJAThorp.com

---

BNT162b2
5.3.6 Cumulative Analysis of Post-authorization Adverse Event Reports

Table 1 below presents the main characteristics of the overall cases.

Table 1. General Overview: Selected Characteristics of All Cases Received During the Reporting Interval

| Characteristics | | Relevant cases (N=42086) |
|---|---|---|
| Gender: | Female | 29914 |
| | Male | 9182 |
| | No Data | 2990 |
| Age range (years): 0.01-107 years Mean = 50.9 years n =34952 | $\leq 17$ | 175[a] |
| | 18-30 | 4953 |
| | 31-50 | 13886 |
| | 51-64 | 7884 |
| | 65-74 | 3098 |
| | $\geq 75$ | 5214 |
| | Unknown | 6876 |
| Case outcome: | Recovered/Recovering | 19582 |
| | Recovered with sequelae | 520 |
| | Not recovered at the time of report | 11361 |
| | Fatal | 1223 |
| | Unknown | 9400 |

Injure-to-Kill Ratio = 33.4

## Do Deaths Occurring after COVID-19 "Vaccinations" Prove Causation?

Some academicians/politicians/self-described "experts" frequently cite this excuse: "Well, the deaths after the COVID-19 'vaccinations' do not prove causation so we should continue using them." This not only demonstrates their scientific and historical ignorance but also their blatant hypocrisy. They refused my public recommendation (and that of others) for a randomized, double-blinded, placebo-controlled trial in the summer of 2020 that would have proven causation in less than one year. Moreover, not even one of the approximately seventy or more vaccines on FDA childhood vaccine schedule has ever been proven safe and effective, with a randomized, double-blinded, placebo-controlled trial with long-term follow-up. Not ONE!

FDA has *never* required proof of causation for removal of drugs—only association. But, considering the formal epidemiologic criteria to prove causation—Bradford Hill Criteria—the deaths from these COVID-19 gene therapy shots fulfill many of these criteria. In otherwise completely healthy individuals, nearly 40 percent of the deaths after the COVID-19 vaccines occurred within forty-eight hours after injection, and a much larger proportion within two weeks after the injection. It is of interest that government institutions, hospitals, and many researchers could assign "non-vaccinated" status to an individual until two weeks after they received the COVID-19 "vaccinations" and this was common practice, especially when considering "efficacy studies."

Consider that in 1976 the swine-flu vaccine was immediately removed from the market for just twenty-six deaths and a spate of cases of Guillain-Barré syndrome—a neurological disorder. Even more impressive on the part of the FDA is that at the turn of the century, they immediately pulled the rotavirus vaccine off the market for several dozen cases of a minor, easily treated bowel complication in young children called intussusception—*there were no deaths*. The government's management of the swine flu and rotavirus vaccines instilled confidence, in that they appeared to be acting in the interests of American citizens. It is incomprehensible that just years after these laudable actions of our government, they would collude with Pfizer and attempt to bury 42,086 casualties including 1,223 deaths in just ten weeks after the rollout of the COVID-19 gene therapy injections.

## Targeting Life Itself

But what happened in early 2021 is even more unconscionable. Not only did the pharma-controlled government attempt to hide this data from

Americans for seventy-five years, but worse, they spent massive amounts of US taxpayers' monies to push the false narrative that these lethal, injurious injections were "safe, effective, and necessary," even in the most vulnerable population of pregnant women, preborns, and newborns.

These vaccines should have been pulled from the market on December 21, 2020, as referenced by Pfizer's 5.3.6 post-market analysis, when there were already over 120 deaths. In fact, these vaccines should never have found a "market."

Now that X (formerly Twitter) is owned by Elon Musk, and people are allowed to post true stories about Covid shots, one readily comes across more horror stories, more unnatural baby deaths.

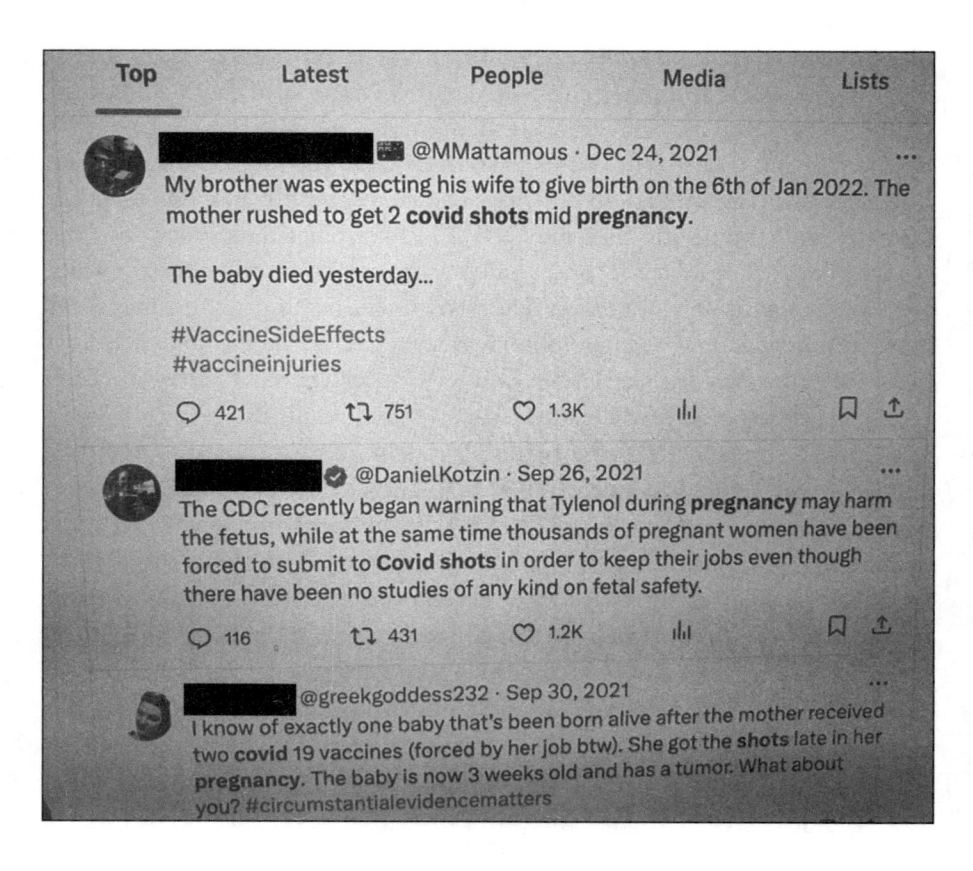

> **Top**  Latest  People  Media  Lists
>
> @MMattamous · Dec 24, 2021
> My brother was expecting his wife to give birth on the 6th of Jan 2022. The mother rushed to get 2 **covid shots** mid **pregnancy**.
>
> The baby died yesterday...
>
> #VaccineSideEffects
> #vaccineinjuries
>
> 💬 421    🔁 751    ❤ 1.3K
>
> @DanielKotzin · Sep 26, 2021
> The CDC recently began warning that Tylenol during **pregnancy** may harm the fetus, while at the same time thousands of pregnant women have been forced to submit to **Covid shots** in order to keep their jobs even though there have been no studies of any kind on fetal safety.
>
> 💬 116    🔁 431    ❤ 1.2K
>
> @greekgoddess232 · Sep 30, 2021
> I know of exactly one baby that's been born alive after the mother received two **covid** 19 vaccines (forced by her job btw). She got the **shots** late in her **pregnancy**. The baby is now 3 weeks old and has a tumor. What about you? #circumstantialevidencematters

# Nurse Bearing Witness

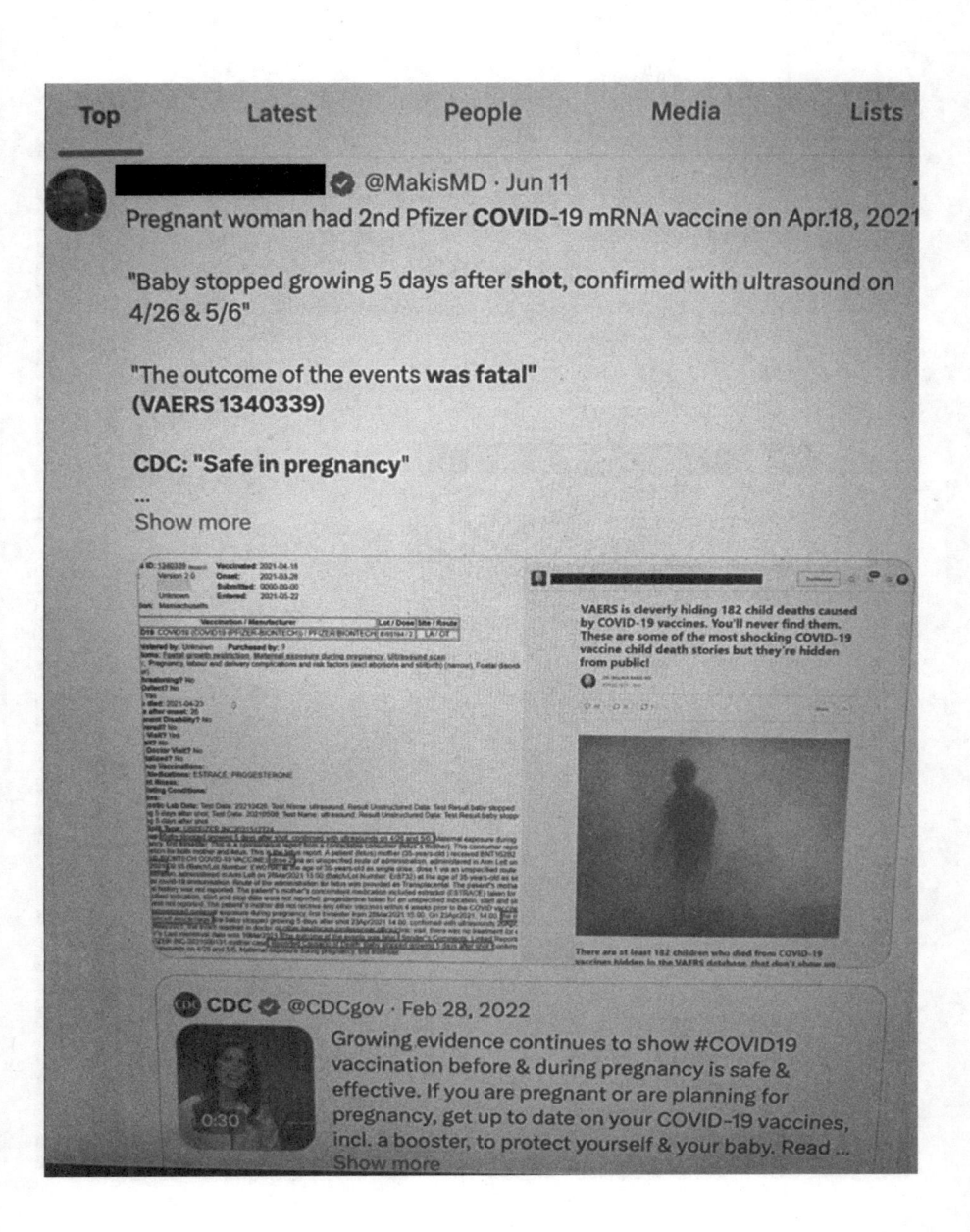

---

| Top | Latest | People | Media | Lists |

**@MakisMD** · Jun 11

Pregnant woman had 2nd Pfizer **COVID**-19 mRNA vaccine on Apr.18, 2021

"Baby stopped growing 5 days after **shot**, confirmed with ultrasound on 4/26 & 5/6"

"The outcome of the events **was fatal**"
(VAERS 1340339)

**CDC: "Safe in pregnancy"**

...
Show more

> VAERS is cleverly hiding 182 child deaths caused by COVID-19 vaccines. You'll never find them. These are some of the most shocking COVID-19 vaccine child death stories but they're hidden from public!

There are at least 182 children who died from COVID-19 vaccines hidden in the VAERS database, that don't show up.

**CDC** ✓ @CDCgov · Feb 28, 2022

Growing evidence continues to show #COVID19 vaccination before & during pregnancy is safe & effective. If you are pregnant or are planning for pregnancy, get up to date on your COVID-19 vaccines, incl. a booster, to protect yourself & your baby. Read ...
Show more

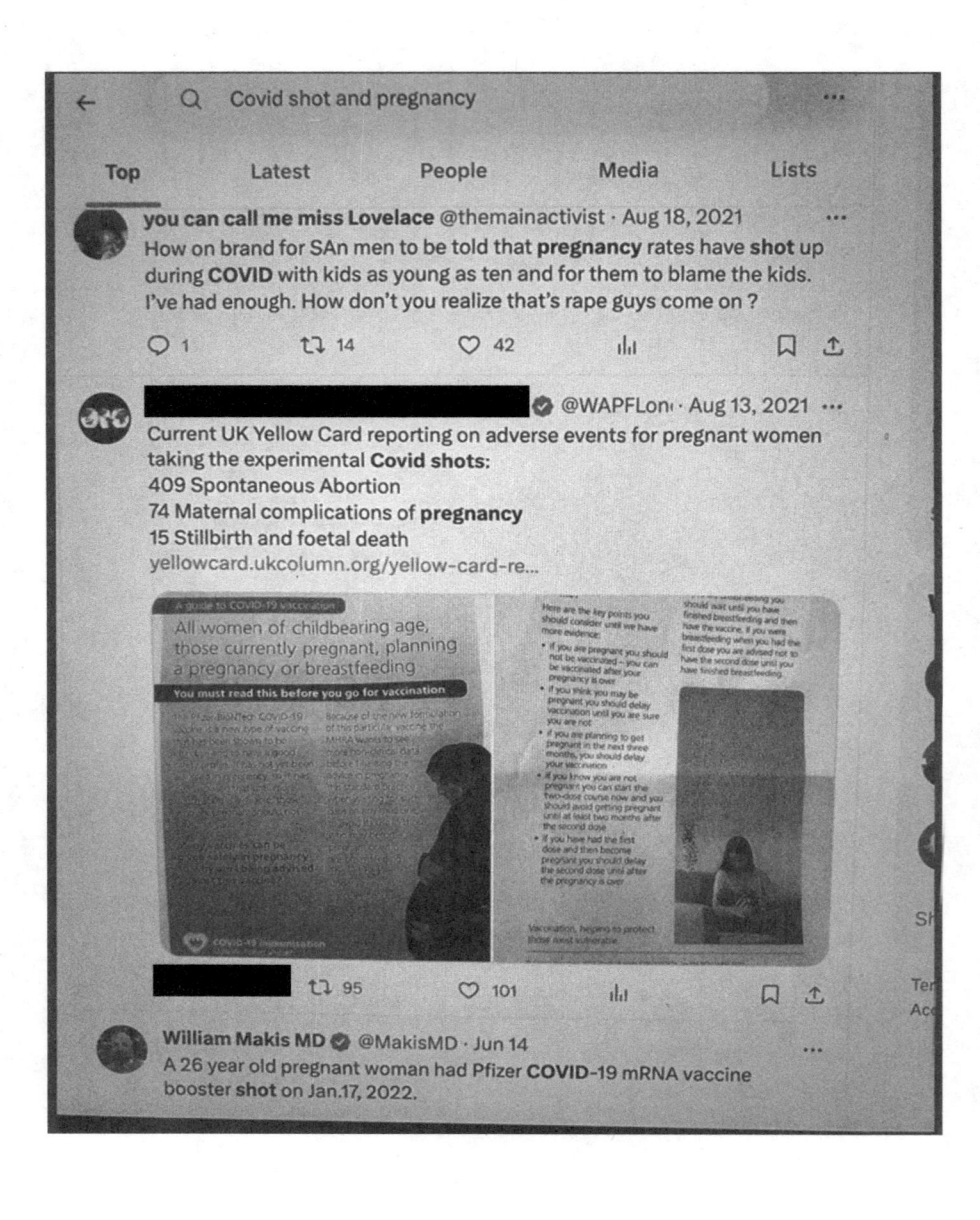

← 🔍 Covid shot and pregnancy ⋯

**Top** | Latest | People | Media | Lists

**you can call me miss Lovelace** @themainactivist · Aug 18, 2021
How on brand for SAn men to be told that **pregnancy** rates have **shot** up during **COVID** with kids as young as ten and for them to blame the kids. I've had enough. How don't you realize that's rape guys come on ?

💬 1      🔁 14      ♡ 42      📊      🔖 ⬆

@WAPFLon· · Aug 13, 2021 ⋯
Current UK Yellow Card reporting on adverse events for pregnant women taking the experimental **Covid shots**:
409 Spontaneous Abortion
74 Maternal complications of **pregnancy**
15 Stillbirth and foetal death
yellowcard.ukcolumn.org/yellow-card-re...

> A guide to COVID-19 vaccination
> All women of childbearing age, those currently pregnant, planning a pregnancy or breastfeeding
> You must read this before you go for vaccination

🔁 95      ♡ 101      📊      🔖 ⬆

**William Makis MD** ✓ @MakisMD · Jun 14
A 26 year old pregnant woman had Pfizer **COVID**-19 mRNA vaccine booster **shot** on Jan.17, 2022.

# Nurse Bearing Witness

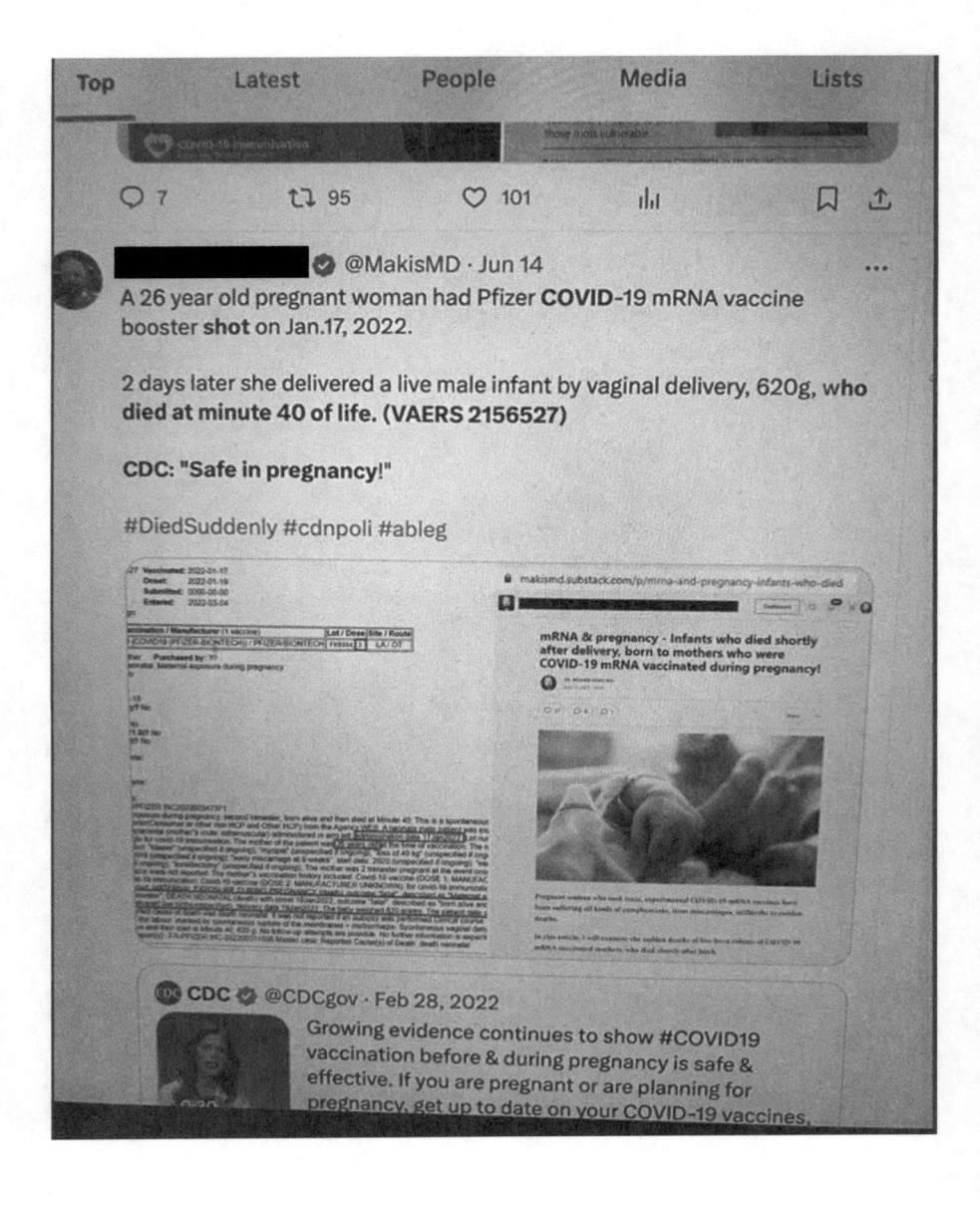

> Top   Latest   People   Media   Lists
>
> ♡ 7   ⟲ 95   ♥ 101   📊   🔖 ⬆
>
> ● @MakisMD · Jun 14
>
> A 26 year old pregnant woman had Pfizer **COVID**-19 mRNA vaccine booster **shot** on Jan.17, 2022.
>
> 2 days later she delivered a live male infant by vaginal delivery, 620g, **who died at minute 40 of life. (VAERS 2156527)**
>
> CDC: "Safe in pregnancy!"
>
> #DiedSuddenly #cdnpoli #ableg
>
> [VAERS report excerpt]   [makismd.substack.com/p/mrna-and-pregnancy-infants-who-died — mRNA & pregnancy - Infants who died shortly after delivery, born to mothers who were COVID-19 mRNA vaccinated during pregnancy!]
>
> CDC ● @CDCgov · Feb 28, 2022
>
> Growing evidence continues to show #COVID19 vaccination before & during pregnancy is safe & effective. If you are pregnant or are planning for pregnancy, get up to date on your COVID-19 vaccines.

> Covid shot and pregnancy
>
> | Top | Latest | People | Media | Lists |
>
> 🇨🇦 **@RobLentz** · May 6, 2022
> Replying to @chelsyhogan
> My best friend and his wife lost their baby at 36 weeks, about 6 weeks after her 2 Pfizer dose
> Doctors recommended getting it so **covid** wouldn't complicate the **pregnancy**
> Official cause was a blood clot in the umbilical cord, of course the Drs rule out the **shot**
>
> 💬 26   🔁 92   ♡ 304
>
> **James Thorp MD** ✓ **@jathorpmfm** · Feb 13
> Pushing **COVID**-19 **Shots** in **Pregnancy**: The Greatest Ethical Breach in the History of Medicine - America Out Loud
>
> americaoutloud.news
>
> 💬 7   🔁 93   ♡ 209   📊 6.3K
>
> **@nathanmhansen** · Jun 1, 2021
> The virus was all about the vaccines from before it started. This is why they want everyone to get the **shots** no matter what: age, **pregnancy**, recovered from **covid**, doesn't matter. To get you in a database and track you and

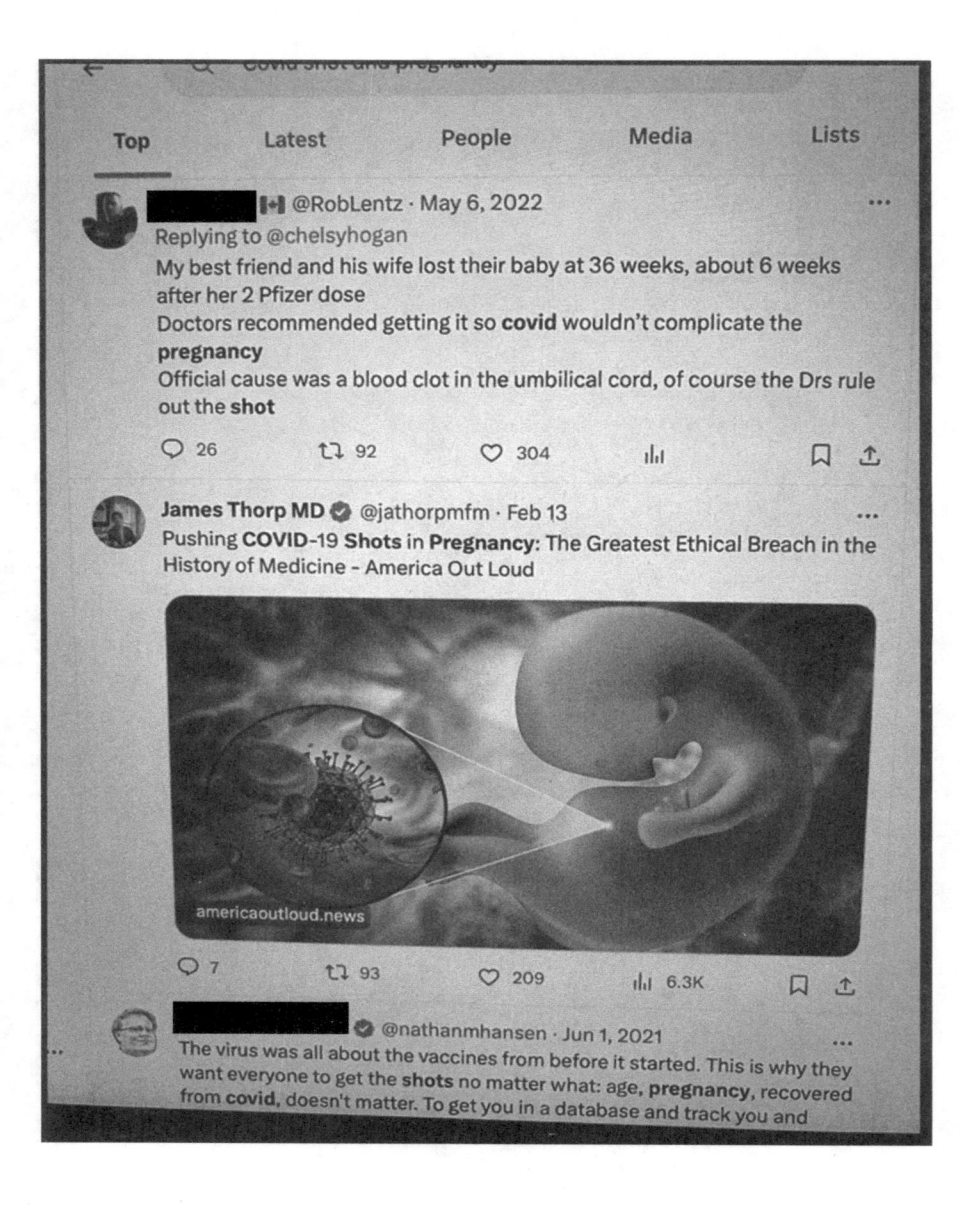

# Nurse Bearing Witness

## "You Hold Them Until They Die"

The popular Substack, *Sage's Newsletter*, on December 29, 2022, published a guest post from a "laborist" who wrote under a pseudonym, an article titled: "You Hold Them Until They Die: It Has Been the Saddest 2–1/2 Years of My 35 Year Career."[3,4]

The line above the article could not have been more clear: "I want people to understand these placental abruptions that lead to dead fetuses. I want them to know some are born alive but too young to resuscitate, so you hold them until they die."

The post is reproduced here, the author used the pseudonym "Brandon is Not Your Bro":

"I was asked to write a segment on Sage's Substack regarding an experience I had working as a laborist at a hospital. I have been posting here regarding some of the cases I have seen and been involved in. Miscarriages, abruptions, stillbirths, heavy menses necessitating blood transfusions, consults for ovarian masses and female malignancies.

"This one case I will never forget.

"Emergency room staff brings a pregnant woman to labor and delivery (18 weeks) bleeding with positive heart tones. Ultrasound at bedside reveals live fetus low in the uterus, placenta is not low lying (no previa) with a large 10 cm clot in the uterus by the placenta.

"Pelvic exam in stirrups reveals dilated cervix and here comes the baby, blood everywhere and baby lands unexpectedly on the gurney thankfully. Alive. Eyes fused, breathing, clamped the cord, baby wrapped in a blanket to the nursery. Parents understand it is too young to resuscitate and refuse to see it at that moment.

"Thankfully the blood was clotting and no signs of DIC. I left with the nurses and kept the senior resident in the room to wait on the placenta. Patient was stable.

"Parents were devastated, deservedly so. The nurse and I stayed with the baby until he took his last breath. Screaming in the room occurred, so I ran back to the patient room; she was in pain due to contractions, and I gave her an additional dose of morphine. I looked in her chart and yes, she was jabbed 3 times. Last jab right before she got pregnant.

"The placenta finally came out with additional cytotec but took a few hours and we did not have to go to the operating room to remove it. The nurse and I prayed over that baby and cried as he passed away.

"Coincidence maybe, but this is occurring at an alarming rate. Usually, the babies are dead, but this little guy wasn't, he tried to live but no chance at 18 weeks. I collect no data. I use trends . . . this is a trend that tells scientists to collect data. I will continue to report the trends I see.

"All we can do is tell people to resist and say no to these jabs. If they choose not to heed the warnings, not much else we can do. I'm seeing a rise in ovarian cancers at the moment. There are still miscarriages but not at the rate it was a year ago.

"Get the word out, please, for the sake of humanity."

## Chapter Three

# Inviting the Devil In: The Machine Model of Biology, Gene Therapy, and the Fall

*Q: The paradigm that we're in now is called what?*
*A: The machine model of biology. That life is a machine, and the elements of the machine are genes and proteins, and that we can reduce the machine to its parts. The idea that we can control life.*
  *I call it an invitation of the devil, this technology. One of the most destructive phenomena in the history of the world.*
    —Richard Strohman, PhD geneticist, and early opponent of "genetic determinism" and "gene therapy," from a 2000 interview with Celia Farber[1]

*The mRNA vaccines are examples for that cell and gene therapy. I always like to say that if we had surveys, two years ago, asking the public, "Would you be willing to take a gene or cell therapy and inject it into your body? We probably would have had a 95 percent refusal rate."*
    —Stefan Oelrich, head of Pharmaceutical Division, Bayer, Germany

## 2020–2021: A Global Terror Campaign and Four Billion Healthy People Injected

It was the fog of war, in 2021, when the "vaccines" arrived, and nobody knew or seemed to care what they were, or how they were made. People had, by then, been so severely traumatized through media mass messaging—an engineered fear of a "virus," possibly from China, endowed with novel and deadly properties, and "lurking" potentially anywhere, surface, air, or skin.

The terror spread like wildfire in the kindling of an already badly fractured, traumatized American people, whose media bashed and scolded them daily, rather than offer them valuable information. The Trump era had normalized this intimate, abusive relationship between the American people and its media. People watched "news" to be terrorized, gaslit, and guilt-tripped.

It worked.

By 2020, most Americans were already demoralized by the time the Covid bomb was dropped upon the entire population.

A supposedly highly infectious, deadly form of "flu" that people were made to fear, like it was some form of weaponized Ebola. This was $57 billion of propaganda at work, generating an untold degree of fear in otherwise rational people. So much so, that they would leave their loved ones to die inside plastic tarps in a nursing home, mask their defenseless infants, or punch a neighbor whose mask wasn't on right—all these behaviors induced by trauma-based mind control disguised as "public health."

Health and Human Services (HHS) had, unbeknownst to the American people, created a parallel economy that paid for a parallel reality: $4.6 trillion all told for the entire Covid revolution including an earmarked $57 billion just for propaganda; a campaign of dreary hopelessness, that saturated people's fields of both vision and hearing, wherever they went, until there was no activity of life left to be enjoyed or experienced as it was BC, "Before Covid."

On TV screens acting as a sort of "pandemic Mussolini" was the seemingly avuncular Anthony Fauci—hedging and posturing, never saying anything clearly, always playing all fears toward the middle, and never reassuring Americans about anything.

The spiky, red ball that was supposed to be a "virus"—but was just some CGI (Computer Generated Imagery)—worked a traumatizing, hypnotic effect on people, as did the posters on every door, every subway corridor, bus, taxi—along with the relentless images of hands being washed, and face masks being affixed properly. These images were absolutely everywhere, pushing all else out of human consciousness.

Nobody, it seemed, thought or spoke of anything but "Covid," and in this crucible, a wild need, even desire, for the "vaccine" was forged.

Having been primed with so much potential for death, messaging about death, fear of death, and guilt about killing others, the vaccines arrived like a heavenly benediction—lifeboats to be deployed on a sinking ship.

This was a time of thunderous propaganda, of overnight political idols, like (then) New York governor, Andrew Cuomo, scowling on nightly news and yelling that we needed to "get shots into arms, and fast."

Soon the news media was awash in stories about people lining up in parking lots and convention centers to get vaccinated, and declaring, with preapproved stickers from the Covid Community Corps, that they had done so; they were relieved, and even proud. They brandished their "I Got My Covid Shots" stickers on clothing and social media. It was a new kind of status—not just getting the shots but flaunting them. Feeling liberated by the shots, octogenarians in New England came out of hospital rooms post-vaccination pumping their fists and crying out: "Free at last!"

This was some propaganda, all right—some next-level mind control.

But what were all these products people were so desperate to be injected with? After all, weren't they just like flu shots?

No—they were brand new, and extra special, extra-technically elaborate, the absolute latest in science, which was to the year 2021, what "peace and love" was to 1968—another massive mind operation, long in the planning.

The double helix was the new peace sign; it would liberate us from our chains, or at least save us from death. And newness was part of the trap. One could not see around the corner, into the past, or from whence it came.

Even the dominant company producing one of the two main "vaccines," Moderna, was brand new. This was not just Moderna's first "vaccine," it was its first-ever product—mRNA.

The shots were cutting-edge, and everything was happening so fast; nobody wanted to alienate their friends, relatives, jobs, or social status by questioning anything. That would be "anti-science," selfish, crazy, and deadly, people were told.

Were the new life-saving shots "gene therapy"? Were they conventional "vaccines"? Nobody knew or cared. The propaganda budget had generated armies of new Covid professionals, who spoke droningly, yet confidently to the public about how "safe and effective" the new vaccines were. But then, in the more serious journals, the "risks" of mRNA were being questioned. None of that, however, would be revealed to the public.

A paper published on PubMed's *National Center for Biotechnology Information,* by French independent researcher, Helene Banoun, on June 22, 2023, made a staggering confession, quite matter of factly, that could not have been made in either 2021 or 2022. Titled, "mRNA: Vaccine or Gene Therapy? The Safety Regulation Issues," it actually admits the shocking truth about the dark events of 2021:

> COVID-19 vaccines were developed and approved rapidly in response to the urgency created by the pandemic. No specific regulations existed at the time they were marketed. The regulatory agencies therefore adapted them as a matter of urgency. Now that the pandemic emergency has passed, it is time to consider the safety issues associated with this rapid approval. The mode of action of COVID-19 mRNA vaccines should classify them as gene therapy products (GTPs), but they have been excluded by regulatory agencies. Some of the tests they have undergone as vaccines have produced non-compliant results in terms of purity, quality and batch homogeneity. **The wide and persistent bio-distribution of mRNAs and their protein products, incompletely studied due to their classification as vaccines, raises safety issues** [emphasis added].[2]

No specific regulations existed at the time they were marketed. But the author seems confident this won't be the case in the future, "outside the context of a pandemic."

Sadly enough, the failure and toxicity of an earlier gene therapy product had already been demonstrated.

## Jesse Gelsinger: A Boy's Shocking Death in a Gene Therapy Trial, a Father's Devastation, and a Crisis in the Field

> *They'd killed animals in the pre-clinical work. It wasn't on the consent form. Everybody failed Jesse, even me. My kid paid the ultimate price.*
>
> —Paul Gelsinger, father of Jesse Gelsinger, eighteen, who died in one of the first human experimental trials for "gene therapy," in 1999
>
> *It's my new career, looking at the limitations of genetic determinism, looking at the shortcomings of the science that I myself practiced for 30 years. Gene Therapy (GT) is in theory something to be looked at,*

*but at the same time we have to say that there isn't a single case of any genetic therapy that has ever worked successfully. Not one.*
—Richard Strohman, interview with Celia Farber, 2000

By the mid to late 1990s, "gene therapy" was massively hyped, despite a total void of any reason to be excited, and every reason to be deeply alarmed. The money incentive was, of course, colossal, and the bought hype enveloped and concealed the early warnings that had been quite clear. Could we just switch out defective genes, the way a car mechanic switches out a broken car part? This was the "big new idea" and promise, that had begun to possess scientists in earnest since the 1953 discovery of the DNA double helix by Rosalind Franklin (who discovered the actual molecule via crystallographic image), and James Watson and Francis Crick, who built the model. To say that this discovery changed our world is an understatement. The Human Genome Project—an international project to identify, map, and sequence all genes in the human genome—was launched in 1990 and ran until 2003.

The gene therapy industrial boom was a strange thing, expressive of the new global cult of "scientism." Genetics was everything—environment (epigenetics) was nothing. Gene therapy would generate unspeakable amounts of funding, despite or perhaps because of its amorphousness. To be against it was to be against progress and all that is good.

Here's an example of what said hype sounded like in 1999:

> *Imagine a world where a person could change his or her genetic structure and redirect the future course of evolution in their child and themselves. Through gene therapy, this is a very real possibility. In the future it will be just as easy to change your physical or mental health as it is to get a flu shot now.*

At that time, did we know that the effects of gene therapy are long lasting and could affect future offspring as well as one's own health?

First discovered in the mid-1970s, researchers were able to isolate certain genes from DNA. During the 1980s, the term gene therapy was coined, propelling further research. The definition of gene therapy is a "technique where the genes causing a defect are themselves substituted by correct genes in the patient to cure a disease."

Unless, of course, something else happens . . .

## A Death That Shook the World

Jesse Gelsinger was eighteen years old when he volunteered for a clinical trial at Pennsylvania State University (Penn State) to test the effect of gene therapy on a rare metabolic disorder called OTC Deficiency. Within hours of being infused with "corrective genes" encased in weakened adenovirus, Jesse suffered multiple organ failure. Days later, his blood almost totally coagulated, swollen beyond recognition. Brain dead, he was taken off life support.

His death, in the year 2000, caused the then-booming field of gene therapy to grind to a sudden, screeching halt. When I (Celia) went to Penn, as my first stop on the interview tour for the *TALK* article, the head of PR said:

> "Not sure what to tell you. We killed an eighteen-year-old kid."

Let me emphasize: Those are words spoken by the head of PR for the medical center where the death happened. I also interviewed Jesse's bereft father, Paul, in depth. It would be twenty years before I would learn that no, gene therapy did not "grind to a screeching halt," as myth would have it. Rather, the scientist at the helm, Dr. James Wilson, who caused Jesse's death in his reckless zeal, was instead, quietly, covertly funded by a $29.4 million grant from GlaxoSmithKline, to keep working on "gene therapy"—right after this happened.

Contrary to popular myth, Gelsinger's death did nothing to stop the beast in its tracks. It became known that a second person—in the same Penn State clinical trial that claimed Jesse's life—became acutely sick and almost died.

None of the scientists who designed the trial anticipated the blood coagulation, jaundice, and organ failure that killed Jesse. He had been so excited to partake in this trial—unknowingly trading his very life by trusting an experiment to correct a genetic defect he could easily have lived with.

An article by Meir Rinde at the Science History Institute recalls the impact of Gelsinger's death:

> Biochemist, Jennifer Doudna, who later discovered the CRISPR-Cas9 gene-editing mechanism, remembers feeling the shock waves as a young researcher, even though her work had nothing to do with gene therapy or any kind of medical research.
>
> "We were all very much aware of what happened there and what a tragedy that was," she said in a recent interview. "That made the whole field of gene

therapy go away, mostly, for at least a decade. Even the term gene therapy became kind of a black label. You didn't want that in your grants. You didn't want to say, 'I'm a gene therapist' or 'I'm working on gene therapy.' It sounded terrible."[3]

Of course, the field eventually rebounded. In the twenty years since Jesse's death, private and public ventures have invested billions of dollars in efforts to cure diseases by altering or replacing our faulty genes.

## Vaccines Reimagined: The Nightmare Unfolds

Messenger RNA gene therapy delivered in lipid nanoparticles, and popularly known as the Covid-19 vaccine, hit the culture like a comet in 2021. There was never any public space in which to have a legitimate discussion about safety testing. From the outset, many doctors and scientists saw this as murderously irresponsible and worthy of professional and personal annihilation.

The lipid nanoparticles (LNPs) are small fat particles that act as a trojan horse carrying the mRNA inside them. LNPs were specifically designed to cross every God-made barrier in the body including that of blood-brain, blood-testes, blood-ovarian, blood-thymus, blood-placental, and all the fetal-blood barriers. This occurs because LNPs are fat-soluble and extremely small in size—one-millionth of an inch. No cell in the body of a pregnant woman and her preborn (fetus) can hide from LNPs. Schädlich and colleagues proved this in 2012; LNPs concentrated in the ovaries in animal models.[4]

The Freedom of Information Act request from Dr. Byram Bridle, of Pfizer's Japanese Biodistribution Study,[5] proved there was widespread distribution to all tissues, including concentration in the ovaries, 118-fold higher in just forty-eight hours after injection. The slope of that rise was so steep that had the animals been sacrificed at ninety-six hours instead of forty-eight hours, the concentration in the ovaries may have reached 1,000-fold. Post-marketing studies have shown that mRNA passes into breast milk and could have adverse effects on breastfed babies. Long-term expression, integration into the genome, transmission to the germline, passage into sperm, embryo/fetal and perinatal toxicity, genotoxicity, and tumorigenicity should be studied in light of the adverse events reported in pharmacovigilance databases. Most authorities now in 2024 agree that the "shedding phenomena" referring to the transmission of toxic substances from a COVID-19 vaccinated person to another person is real. Exactly what is being shed—be it mRNA, lipid nano particles, spike protein, or other substances—is unknown.

## Was It a Government-Sponsored Bioweapon as Some Claimed?

It sounds crazy to even ask. In our old "normal" world, it would be insane to think that our own government would aid and abet the creation of a laboratory bioweapon virus that would trigger the declaration of a suspect global pandemic, and would lead to worldwide economic destruction, death, and the production of a "vaccine" with a shocking kill rate (and adverse events) across all age groups.

Most ominously—a bioweapon that keeps on attacking the body long after it's been administered.

Furthermore, the vaccine's effects on human reproduction—what should be a top priority for us humans as a species—are deeply disturbing. Even worse, new evidence suggests that the "vaccine" implants permanent DNA codes in the bodies of the injected, which means whatever adverse effects that DNA causes will be passed on forever to future generations.

Keep in mind as you read the rest of this chapter, that while Pfizer was testing the "vaccine" for emergency authorization approval, it recommended that pregnant and lactating women *not* be included in the testing. They gave no reason, but there had to have been some concern about the effects of certain components of their concoction on mothers and their babies during the most delicate time of creating life.

## Emergency Use Authorization, Pfizer, and Informed Consent

Any concerns that Pfizer may have had regarding the safety of their experimental gene product vanished after "Emergency Use Authorization" (EUA) was granted. EUA's are declared when it is found that there are no other effective treatments available on the market. Now, it was up to those administering the shots to convey cautionary information to people before they injected them so that those people could make an informed decision about whether or not to go through with it. This is called "informed consent." By and large, this didn't happen. In fact, even though the Emergency Use Authorization excluded pregnant and lactating women, doctors were encouraging that cohort to take the shot.

To make matters even more surreal, the inserts on the vaccine packaging—listing ingredients, contraindications, and side-effects—were intentionally left blank. Prescription medications are always accompanied by an insert, both sides filled with detailed information about relevant information on the drug—including every possible known side-effect. Except, for the COVID-19 vaccines, the package inserts were left blank.

I demanded the pharmacist at Walmart Pharmacy, (334 Gulf Breeze Pkwy, Gulf Breeze, Florida 32561), pull the insert out of the package containing the COVID-19 vaccine in early 2022. She was stunned by what she saw:

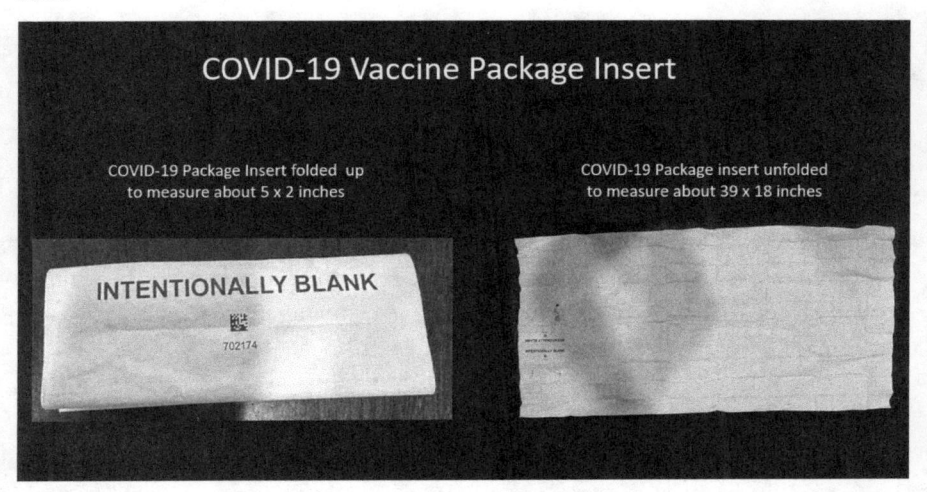

**COVID-19 Vaccine Package Insert**

COVID-19 Package Insert folded up to measure about 5 x 2 inches

COVID-19 Package insert unfolded to measure about 39 x 18 inches

INTENTIONALLY BLANK
702174

In addition to hiding important data from the public, CVS and Walgreen's pharmacies were paid over $4 billion US tax dollars to push the lethal vaccines, gaslight physicians, and *not* fill prescriptions for doctors who had ordered ivermectin and hydroxychloroquine for their patients to treat Covid.[6]

Had the care providers been operating in a conscientious and legal manner, here's what they would have told all pregnant patients about these products: that the vaccine was experimental and not licensed by the FDA; that it was authorized for emergency use; that the authorization had excluded pregnant women; that pregnant women were excluded from the vaccine clinical trials; that the vaccine doesn't provide immunity or stop transmission of the virus; that the vaccine lacks durability; that it doesn't stay at the injection site but instead travels through the bloodstream; and that it poses serious known and unknown safety risks to the mother and baby, including fetal death and congenital abnormalities.

Not only did most obstetricians not do this, they were financially incentivized and pressured by their professional organizations, the FDA, and the CDC to encourage pregnant and lactating mothers to get the shot. I find this unconscionable.

Any mother's ears would have perked up at the "serious known and unknown safety risks" part of the informed consent talk if they'd heard it.

Any mother would likely want to ask what was in the vaccine and what those risks were.

Those two questions would be much harder to answer if certain extraordinary events hadn't occurred in 2022.

First, attorney Aaron Siri managed to pry fifty-five thousand Pfizer documents out of the Food and Drug Administration, who had asked the court to keep them under wraps for seventy-five years. Then, Dr. Naomi Wolf, founder and CEO of DailyClout.io, a website self-described as "devoted to civic transparency," took on the huge task of scrutinizing all the documents. Dr. Wolf gathered a team of more than three thousand relevant experts from around the world to assist in the research. The result is the *Pfizer Documents Analysis Report*, a series of forty-six reports exposing what is described in the report's foreword as "what may be a massive crime against humanity."

The foreword promises the reader an eyeful:

> You will see that Pfizer knew, as it appears, that the mRNA vaccines did not work . . . you will see that Pfizer and the FDA knew the injections damaged the hearts of minors—and yet waited months to inform the public. You will see that Pfizer sought to hire over a thousand new staffers simply to manage the flood of "adverse events" reports they were receiving and that they anticipated receiving . . . you will see neurological events, cardiac events, strokes, brain hemorrhages, blood clots, lung clots, and leg clots on a massive scale. . . .

Then came these words that resonated with my own traumatic experience, validating what I knew, but horrifying to me all the same:

> Most seriously of all, you will see a 360-degree attack on human reproductive capability; with harms to sperm count, testes, sperm motility; harms to ovaries, menstrual cycles, placentas; you will see that over 80 percent of pregnancies in one section of the Pfizer documents ended in spontaneous abortion or miscarriage. You will see that 72 percent of the adverse events in one section of the documents were in women, and that 16 percent of those were "reproductive disorders," in Pfizer's own words. You will see a dozen or more names for the ruination of the menstrual cycles of women and teenage girls. You will see that Pfizer defined "exposure" to the mRNA vaccine as including skin contact, inhalation and sexual contact, especially at the point of conception. [7]

In an *America Out Loud* publication, my wife, Maggie Thorp JD, and I published an article entitled, "A call for Immediate Moratorium on the use of COVID-19 Vaccines in pregnant women."[8] I include a summary here, from my X (formerly Twitter) account:

---

How The Lies About Reproductive Concerns Collapsed in Less Than 3 Years
Maggie Thorp JD & Jim Thorp MD. "A call for Immediate Moratorium on the use of COVID-19 Vaccines in pregnant women". March 5, 2024. America Out Loud

2021 - COVID-19 "vaccines" remain in the arm

2022 – Alden et al demonstrate the vax mRNA is reverse transcribed into the human liver cell in vitro

2022 – Hanna et al demonstrate intact vax mRNA is incorporated into lysosomes/exosomes and excreted into human breast milk potentially integrating into the breastfed infant

2023 – Hanna et al: confirms their 2022 report in yet a separate study in 2023

2024 – Lin et al:  Vaxx-mRNA crosses the placenta into fetal blood, incorporates into the placenta and uterine lining (decidua) with heavy spike protein production in placenta AND uterine decidua

2024 - Mikolaj Raszek provides preliminary evidence that C19 vax may be permanently reverse-transcribed into the human DNA genome

James A Thorp MD on X @jathorpMFM

---

In the article, we described the Lin study, published January 31, 2024, by the *American Journal of Obstetrics & Gynecology* indicating that the COVID-19 mRNA vaccine is not, in fact, localized to the injection site, but rather can spread systematically to the placenta and fetal blood.[9] For context, the *American Journal of Obstetrics and Gynecology (AJOG)* is a peer-reviewed journal of obstetrics and gynecology. Popularly called the *"Gray Journal," AJOG* is considered by most to be the pinnacle of journals in the specialty of Ob-Gyn. Since 1920, *AJOG* has existed as a continuation of the *American Journal of Obstetrics and Diseases of Women and Children,* which began publishing in 1868.

Lin found that COVID-19 vaccine mRNA was detected in the placentas of two pregnant mothers who had been vaccinated with Pfizer's mRNA COVID-19 vaccine shortly before delivery. The study also found that spike protein expression was detected in the placental tissue of the earlier-in-time vaccinated mother, demonstrating bioactivity of the COVID-19 vaccine mRNA after reaching the placenta. Also of great concern, vaccine mRNA was detected in the fetal blood of the only patient sampled, documenting a 100 percent rate of transmission in their study. The implications of these research findings are profound. They indicate that COVID-19 mRNA vaccines penetrate the fetal-placental barrier and reach the intrauterine environment, where the mRNA can then be translated into spike protein and expressed in the placental tissue.

Shockingly, the now documented transplacental transmission of COVID-19 vaccine mRNA did not appear to alarm the pro-vaccine New York research group, whose study was funded by the Eunice Kennedy Shriver National Institute of Child Health and Human Development (NICHD), which is one of the institutes of the National Institutes of Health (NIH). This is ironic, given that transplacental delivery of the mRNA vaccines was previously debunked as false and derided as "misinformation" in 2021. For example, in August of 2021, the fact-checking organization *AFP* accused Dr. Ryan Cole of spreading misinformation and called his claims false, when he asserted in a social media post that, after receiving a COVID-19 mRNA vaccine, "the spike protein doesn't just stay in the deltoid, the spike circulates in your blood and lands in multiple organs."[10] New York–based news organization the Associated Press also ran a fact-checking piece in June 2021, calling the claim that the "mRNA vaccines don't stay in the shoulder muscle" false, and instead alleging that the vaccines are "mostly concentrated at the site of the injection."[11]

Perhaps fueled by so-called "misinformation" claims that the COVID-19 mRNA could be distributed throughout the body, and perhaps pressured to quell such concerns, authors from University of California, San Francisco (UCSF) and the San Francisco J. Gladstone Institute assessed the possibility of systematic biodistribution in pregnancy in an earlier 2022 study. Entitled "Evaluation of transplacental transfer of mRNA vaccine products and functional antibodies during pregnancy and infancy," this study, published July 30, 2022, in *Nature Communications*,[12] looked for transplacental transfer of mRNA vaccines products and SARS-Cov-2 antibodies in a cohort of twenty vaccinated mothers during pregnancy. Authors of the study claimed to have found no evidence that mRNA vaccine products were distributed to the maternal blood, placenta tissue, or fetal blood. They concluded that while transplacental transfer of protective *antibodies* can occur, products of mRNA vaccines are not transferred to the fetus during pregnancy.

Notably, UCSF and the J. Gladstone Institute (institutions associated with the study authors) have been long-standing recipients of federal money, having received tens of millions of federal dollars, and in the case of UCSF, hundreds of millions of federal dollars. In fact, UCSF proudly boasts on its website that UCSF has received more funding from the NIH than any other public university for seventeen years in a row. Perhaps it is no surprise that they attempted to weigh in early, in what would temporarily squelch concerns that transplacental transmission of mRNA vaccines was not only plausible, but actually occurring.

Importantly, as Maggie and I point out in the reference cited above, the UCSF and J. Gladstone Institute author conclusions, with regard to transplacental transmission of mRNA, which were published in 2022, *have now been directly contradicted by the findings in the latest research reported in the AJOG Lin article cited above.* These two polar-opposite and contradictory narratives, rendered within such a short period of time, shine a light on the fallacy of "scientific evidence-based medicine" as we know it today. **As the government's narrative changes, so changes the so-called "science."**

The Lin study published in *AJOG* on January 31, 2024, documenting the transplacental transmission of mRNA from the COVID-19 vaccines, demonstrates that the cart went before the horse in urging pregnant women to unhesitatingly take the COVID-19 vaccine.[13] Yet, the *American College of Obstetricians and Gynecologists* (ACOG) claimed that the COVID-19 mRNA vaccinations are safe and effective, recommending them for those who are pregnant, postpartum, breastfeeding, or planning on becoming pregnant.

Consider this: even the Lin AJOG study's pro-vaccine mRNA authors—who at times seem to be speaking out of both sides of their mouth (perhaps in an effort to please their funding sources and journal editors)—admit that the mRNA vaccines pose plausible risks to the fetus. The authors state:

> ***Although gene therapy, particularly mRNA-based treatments, shows promise, research on its perinatal delivery is still emerging.*** Prenatal therapy can be advantageous, because it offers early disease intervention and reduced immunogenicity. In experiments with pregnant rats, LNPs successfully delivered various mRNAs, including one potentially useful for treating fetal anemia. ***Although introducing mRNA to the fetus may potentially pose plausible risks,*** it may also have biologically plausible benefits [emphasis added].[14]

Two assertions made by the study authors should give those pushing the mRNA COVID-19 vaccines in pregnant women concern. First, as the study authors essentially admit, a sobering fact remains—*no one knows the long-term risks the COVID-19 mRNA vaccines might have on the offspring of pregnant mothers.* Second, despite their obvious pro-vaccine position, the *AJOG* authors readily admit that the mRNA products are "gene therapy"—an assertion also previously debunked as false by the US government federal money recipient and mouthpiece, Associated Press.

## The Ontogeny of Lies about the COVID-19 Vaccines

When we consider the litany of misrepresentations about the COVID-19 vaccines that have been perpetrated on the American public since their roll-out in December 2020, concerns about reverse transcription and harm to the genome take on added significance. Misrepresentations have emanated not only from our government agencies, but also (and perhaps more concerningly), from a myriad of "trusted" sources—such as medical journals, hospitals, nurses, physicians, and government health administrators. Consider the following examples, which include just a few of the misrepresentations told to the American public during the pandemic:

- Americans were initially told that nobody would be mandated to take the experimental COVID-19 injections—which were rebranded as "vaccines" to make them more palatable to the masses—including pregnant women, who were not included in the pre-clinical trials.
- Hydroxychloroquine was villainized as unsafe in a *Lancet* journal article by lead author Mandeep Mehra in May 2020.[15] Yet, the CDC has previously deemed it safe enough for pregnant women, nursing mothers, and children of all ages, as shown in a CDC directive prepared for malaria prevention for overseas travelers. Although the *Lancet* article was retracted five months after publication, the damage had been done—many people were more afraid of taking hydroxychloroquine than they were of contracting COVID-19.
- In a June 2021 article published in the *New England Journal of Medicine*, lead author Tom Shimabukuro claimed that there were no safety concerns with the COVID-19 vaccines during pregnancy, despite Pfizer's post-marketing report documenting it to be the deadliest and most injurious medical intervention ever rolled out.[16] Upon careful review, the study authors—all of whom had conflicts of interest as federal employees (including being part of the CDC V-safe COVID-19 Pregnancy Registry Team)—either made gross errors or relied on statistical sleight-of-hand to reduce a miscarriage rate documented in the article as 82 percent (104/127) to a more palatable (yet still too high) miscarriage rate of 12.6 percent (104/827).
- In 2020 and 2021, the pharmaceutical-industrial complex claimed that the mRNA from the COVID-vaccines remained localized in the deltoid muscle of the arm. Concerns that the mRNA could potentially be reversed transcribed and incorporated permanently

into the human genome were downplayed as misinformation. But alarmingly, lead author Markus Aldén and colleagues demonstrated in February 2022 that this very phenomenon occurred in human liver cells *in vitro*.[17]

- Even more alarming is that two separate studies, published in 2022[18] and in 2023[19] by lead author Nazeeh Hanna, demonstrated that intact vaccine mRNA is excreted into human breast milk. These findings opened up the very real possibility that lipid nanoparticles could be distributed to every cell in the body, repackaged in the cytoplasm, resulting in mRNA excreted via liposomes or exosomes to be redistributed to potentially all other exocrine glands in the body. The vaccine mRNA could then be excreted by exhalation, sweat, cervical-vaginal secretions, and prostatic secretions. This, in addition to the spike protein, could explain the shedding phenomenon.

In summary, the evolution of mRNA vaccine "misinformation" from 2020, to what has been documented as true in 2024, has been astounding. Consider—just four years after Americans were falsely told the mRNA in the COVID-19 vaccines remains localized in the deltoid muscle, experts now credibly suggest that the vaccine mRNA could be permanently reverse-transcribed into the human genome. Research by Lin published by *AJOG* this year (referenced above) has demonstrated that mRNA from the COVID-19 vaccines can cross the placenta into fetal blood, and also enter placental tissue and produce spike protein. This brings heightened urgency to the question of whether COVID-19 mRNA is being reverse-transcribed into the human genome.

## Are the mRNA Vaccines a Contributor to Maternal Morbidity and Mortality?

Also of great concern, the Lin *AJOG* article observed a "notably high signal" of vaccine mRNA in the decidua (which is the lining of the uterus closest to the fetus). Concentrated mRNA in decidual tissues can be translated into high concentrations of spike protein, which would plausibly contribute to a myriad of adverse effects on human reproductive function, including not only menstrual abnormalities (which has been previously documented) but also severe bleeding in pregnancy and in the post-partum period—a dangerous condition that can even lead to the death of the mother. When one considers that the CDC reports that maternal mortality had skyrocketed in 2021, given the notably high signal of vaccine mRNA in the decidua reported by the *AJOG* authors, the question of whether the mRNA vaccines have been a contributor must be examined.

## Do mRNA Vaccines Risk Permanent Alteration of the Human Genome?

Finally, in February 2024, geneticist Mikolaj Raszek, PhD of Merogenomics, reviewed preliminary data suggesting that the vaccine mRNA could be permanently reverse-transcribed into the human genome (DNA) *in vivo*.[20] While the evidence is not a smoking gun, his preliminary findings have potentially catastrophic implications. If the vaccine mRNA is reverse transcribed into the DNA of the "germ cells" (sperm and/or ova), there could be permanent genomic alterations in the DNA of future generations. With the rapid evolution of designer polymerase chain reaction (PCR) sequencing, it is likely that Raszek's preliminary data will be more thoroughly evaluated in the near future. If history is a teacher, given the ontogeny of lies told to the American public over the pandemic years, permanent integration of the vaccine mRNA into the human genome could be established, which may well be permanently incorporated into the genome of future generations.

The "science" of all this becomes yet another tomb; an endless attempt to document, and quantify, something unimaginable. Those who would nitpick, or cling to the fantasies of "scientism"—which bedevil the Truth and Freedom side as well—should let it all go at a certain point, and listen, attentively, to a single man on the ground, at the scene of the crime.

A man like Harry Fisher.

Chapter Four

# Interview: A Paramedic Who Became an Enemy of the Covid State

*There's no question of heroism in all this. It's a matter of common decency.*

—Albert Camus, *The Plague*

Harry Fisher is a traveling paramedic who has emerged as one of the most raw, trustworthy, and targeted voices in the Covid truth battlefield. When Covid descended, he saw no actual "Covid," but when the shots were rolled out, his life was turned upside down as if by an earthquake.

It began with emergency calls over people dropping dead at vaccination stations—two deaths in two weeks at a single clinic that Harry witnessed. Soon his every working moment was a surreal, blood-soaked nightmare he could never awaken from. Pregnant women arriving in droves at hospitals, miscarrying, nearly bleeding to death—once it was nine in a single shift. When an MD turned to Harry—himself unvaccinated—and said: "We're witnessing a genocide," he stopped wondering if he was crazy. He trusted what he was seeing, and began to record and post short, spoken testimonials on social media.

TikTok not only banned him but banned an additional fifty accounts each time he tried to get back on. Even X torpedoed his account—later restored. He's had his license threatened, his life threatened, his car tampered with, and every single day he is attacked by swarms of online trolls, viciously trying to discredit him. A devout Christian and single father, he spoke to us for this book, after an all-night journey back home from a job he was called to work in rural Alaska.

This transcript is Celia Farber asking the questions, Harry answering, and Dr. Jim Thorp interjecting his medical observations. It was recorded in May 2024:

**Q:** Could you describe for us the first time you realized in your work, in your life, that there was a catastrophe unfolding with the Covid shots? When did it hit you?
**A:** It started hitting me when I did CPR on a Pfizer line. That was a big red flag. The nurse came over and said it was the second person that had died in two weeks in that same clinic—those makeshift shock clinics that they would put up. That was a big red flag.

And then I just started getting calls that were abnormal. I've been doing this work since 1997, you can see when patterns change. It started with younger chest pain calls, *way* more seizures. I had one hospital shift where there were nine miscarriages in one shift. Spontaneous abortion miscarriages, all of that. It was like a process of coping and really just processing the information because it's not like you want to immediately think that your government could do this to you, or some big evil bad company could do this.

It's just so bad. And it takes a while to actually process that information. We don't look at evil like that, you know, as Americans.

It's things you see in history books when you're thinking of like Hitler or something bad, that's happened in the past. So, whenever you're seeing it occur, you know it's happening, you know bad things are occurring, but it really takes a while for you to cope with the extent of it. So, I would say it was a process over months, a bunch of red flags going up. And then finally realizing how bad it was.

A turning point for me was when a doctor came up to me and said, "We're experiencing genocide." That was all she said.

And that right there was when I decided to go out and make a TikTok video. I didn't know what else to do.

Supervisors already knew what I'd seen. People I've already told, you know, face to face. So, I was like, "Okay, I'm going to make a video."

And I made a video. I was shocked at how many views it got. Over a million views, within twenty-four hours.

But the comments, thousands and thousands of people. "I lost my brother," "I lost my uncle," "I lost my aunt," "I lost my mom." And then they just deleted all of it—TikTok just deleted it all.

"Terroristic activity" is what they said. All my original video said was: "I'm a paramedic. I did CPR. I lost two patients within two weeks." Boom, short to the point, that's it.

And they called that "terroristic activity."

**Q:** Was that your first TikTok?
**A:** That would be my first medical TikTok.
**Q:** Now this colleague who said, "We're experiencing a genocide," who was that?
**A:** ER doctor.
**Q:** Has that ER doctor come out in any way or gone public?
**A:** I don't believe so. We were on a reservation, and she'd been working there for years, and she believed it was genocide. Enough to where she said she was retiring, going to her ranch in North Dakota and shutting the door. She was done. She knew the writing on the wall, and she got out, from what I hear.
**Q:** What was your opinion of vaccinations before all this happened?
**A:** I was in. I was pretty brainwashed just like everyone else, thought vaccines were safe. I was pretty indoctrinated. Like standard. Yeah.

It really hit me that other vaccines are bad because I'd been seeing this hurt people. I know a senator who got ahold of me through TikTok. You know, what's sad is that even senators—they're having to watch TikTok videos to try to find information. We were all so censored and a senator invited me to this dinner. I went to go to dinner with him and I was sitting at the table, and I was going to talk to some people about what I'd seen, you know, with the Pfizer shot and the other vaccines, just the other COVID-19 vaccines. And I was sitting with this couple. Their kid had a vaccine injury—the MMR paralyzed their child.

And then I started to realize, "Oh God, if they would lie to you about something like this, if they're going to censor all of us over this, the system's going to call you a terrorist just for trying to warn people, then how deep do the lies really go?" I had to take a step out of the box that they put me in, you know, the indoctrinated box. If the mainstream news is saying it's safe and effective, I'm pretty much going to think the opposite at this point.

**Q:** You said you've been a traveling contract EMT since 1997, paramedic since 2012?
**A:** Yes. I was in the army air force as a medic in both branches. And then after that, I was ambulance. And then when I got paramedic, I started doing contracts right as Covid hit. Basically, that's when I started picking up contracts.
**Q:** You made a video once about the nine miscarriages in one shift.
**A:** Yeah. I'll never forget that one. It was really bad. So, imagine the packed ER. And it was during that time, a time where we were having difficulty finding beds. We were having to turn a portion of our ER into like an ICU or, you know, longer-term care. So a lot of these ladies who were having miscarriages, they're stuck sitting in bloody chairs. It was really grotesque.
**Q:** And how far along were they?
**A:** They ranged anything from, you know . . . anywhere from like ten, twelve weeks to . . . later. They were a wide range, wide range.
**Q:** Are you able to tell us what city or state?
**A:** That one was in Oklahoma and North Dakota. I saw similar things in both states, North Dakota and Oklahoma.
**Q:** Now, was there any talk on the floor with your colleagues about that? Did anybody say that the vaccines might be the cause?
**A:** One of the PAs that was sitting next to me, this is a quote from a PA who was sitting there whenever I was. I said: "This is insane." I was basically just venting about what we were seeing. And I was very vocal about the vaccines. Whenever I started seeing bad things happen, I'm going to talk about it, which didn't make me very popular.

But the PA actually looked at me and said, "I'm putting my head in the sand." So, whenever I say "they're putting their head in the sand," I've literally heard PA's say that.
**Dr. Thorp:** At least that person wasn't in denial over how bad it was.
**Q:** Have you had threats on your life or threats against your work or any kind of intimidation?
**A:** Oh, a hundred percent. Since I started speaking out, I've had my car broken into. They drew little pictures inside my car for intimidation. Instead of taking anything, they just broke in and drew these little emblems in my car. Very strange. Yeah.

I've had them come after my license. They've called the national board and tried to get my license taken away from me.

They, they come after me online all the time. I mean . . . you don't have to look very far on my pages, and you'll see them just coming after me on there. I used to block them and now I just mute them.

I know they're trying to actively get me because they've, they've gotten me suspended before and they suspended me for impersonation. X told me that I was impersonating myself and it took a couple months for them to realize, oh, he's got a blue check mark. He's obviously not impersonating himself.

So, they gave me my account back. But instead of blocking those evil pharma people, I just let them talk. That way people can at least see how bad they are.

If I block them, they're just, they're going to still be talking, you know, somewhere.

*Res ipsa loquitur*, right? The situation speaks for itself. . . . you would think it couldn't get any more obvious at this point.

But people are still grabbing ahold of that denial and running with it as long as they can. It's wild. Like I've said a couple of times now, I mean, I liken it to abused people. They really remind me of that abused spouse that's been beaten and beaten, and they don't want to report it. They don't want to believe it. They don't want to admit it.

And they'll just get mad at you. You bring it up too much. They're just going to come at you and try to shoot the messenger.

It's wild. It's like Munchausen by proxy meets Stockholm Syndrome.

**Q:** If you can just comb through your mind, Harry, specifically about the mothers and the pregnant women. How dangerous are these shots to pregnant women and what kind of things have you seen?

**A:** I would say they're very dangerous to everyone, but especially pregnant women. If you're taking the shots as a pregnant woman, you're playing Russian roulette with your baby's life. I can't even tell you how many miscarriages I've witnessed. I did notice it was just a lot more bleeding than a miscarriage would have been before. I don't know why. I'm not a physician, but I can tell you they're bleeding much more. There's a lot of blood. It's just a whole lot of blood.

I have a lot of bad dreams constantly. Recurring nightmares. It's . . . difficult. And then the cleanup . . .

I mean, dead babies are not something you ever get used to. If you talk to someone who says that it doesn't bother them, there's, there's either something wrong with those people or they're lying to you. I mean, I don't know anybody that dead children doesn't actually affect them.

You try to not think about things like that, as much as possible.

People cope in different ways. I've coped with excessive alcohol use.

I've coped with, you know, you self-medicate for a long time, especially as a paramedic. A lot of us self-medicate with some doing hard drugs. Mine was alcohol.

I'm currently trying to consume no alcohol at all, which is, well it's a task. Like I'm trying to kick it all and just have my coping mechanism be God.

It's, it's very difficult because you want to sleep. And I would drink when I'm off work and I, . . . I'll go for weeks at a time, and just work. And then whenever I get home and I'm, I don't have my kids, and it's just me staring at a wall, thinking about all the ghosts that are in my head, I just want to go to sleep.

**Q:** I understand. So, back in 2020, the first chapter of course, where everybody was brainwashed to think that people were dropping dead in the hospitals. Did you see that from Covid before the shots?
**A:** No, especially not children's hospitals. Children's hospitals got completely slow.

They were, the parents were too scared to go to the hospital. They weren't in the schools. You saw this wave of just sickness, not really Covid, just all the sickness that used to come in steady waves, you know, cause people are building up their immune system as kids.

And then suddenly they, they keep them from building their immune system and seemingly throw them all out and you're going to get this tidal wave of sick kids. And we, we saw that, and it passed. But I didn't see anybody like . . . massive . . . not until after the shots rolled out. I mean, that's when I saw the dying *suddenly*. I'll give you an example. I saw a young man in his early twenties come in for chest pain.

I talked to him. He had two Covid shots. They did a workup, drew some blood, did an EKG, but that's all. Twenty-year-olds, you know, didn't typically have heart attacks. And if the EKG looked good, that was it, he was sent home.

# Interview

He came back in an hour later saying the chest pain was still there—abdominal to chest pain—and they still dismissed him. And he died, suddenly died at the door of our ER. We couldn't get him back. He ended up having an aortic dissection.

I mean, I've seen more aortic dissections than I've ever seen in my career since the rollout of these vaccines. Those aren't anything to play with. I've seen a lot of those.

**Q:** Could you tell me what an aortic dissection is?

**A:** The tearing where you start bleeding internally, whenever it actually erupts or, you know, you, you have massive amounts of bleeding inside.

And I've never seen one that actually ruptured that we could actually do anything about. Like, I mean, the more CPR, it seems that I do on them, the more they bleed out.

**Jim interjects:** What it is . . . is the spike protein, of course, which attaches to every cell that lines the blood vessel, including the large arteries. And so, it causes inflammation, and that inflammation will then go through to the cell that's closest to the blood and go into the deeper portions, including the separate tissues and the muscle. And then because of the inflammation throughout the vessel wall, it will rupture.

And then those deeper tissues will separate in the vessel wall and start hemorrhaging internally and then there's bleeding into that whole space. So it, it dissects the tissue due to the inflammation.

**Q:** Thank you, Jim. Harry, that young man, twenty years old in the so-called normal world before these shots, did you ever see a twenty-year-old with anything like that before the shots rolled out?

**A:** No, I didn't see twenty-year-olds, you know, collapsing from aortic dissections. I never witnessed it until after the shots. And then I started seeing it more and more.

**Q:** And you also have to interface with the parents' grief every time, right? This is also part of your work.

**A:** Yeah. Yeah. It's difficult, especially when the parents are in denial. I went out on a call, a twelve-year-old girl had a stroke, playing kickball. I've seen plenty of bad stuff, but this one . . .

The mom was an RN and the mom made it to the field. We were at the school, and she got in the ambulance with me. The mom was in such denial that she was saying, "Can it be her blood sugar?" And I'm like, "No, she's paralyzed on one side of her body. She's slurring her speech. She had a thunderclap headache. These are stroke symptoms. You know, you're a nurse."

"Uh, well, is it, is it her sugar? I'll check. I'll check her sugar." Check the sugar. "No, it's not her sugar, ma'am."

"Uh, has she taken any medications?"

"Oh, no, no new medications."

"Okay. Has she taken any of those Covid shots?"

"Well, what do you mean, the COVID-19 shots? Well, yeah, she just had her second Pfizer like a couple of weeks ago, but I mean, can't be *that.*"

They're in total denial. And that's a nurse.

It's hard to talk to them sometimes. And then some patients, you know, they know full well, especially nowadays, a lot of them are starting to realize that the shots are what's causing it all. So, it's changing. It's just, it's been a slow, horrible process.

I'll keep working until I can't anymore. I don't know how, as long as I don't, you know, give away HIPAA information, as long as I don't do anything that hurts patients. What they're trying to do is say that I'm a danger.

You know, the biggest thing that they try to say is I'm dangerous for patients because of my beliefs, that I'm crazy for the way that I believe. And if we lose this battle, if we lose this big battle, information battle, then they'll eventually say that me and doc and you, we're all crazy. And then the rules change whenever, you know, they deem you actually insane, especially like, I guess a paramedic, if I believe you're a threat to yourself or to other people . . . I can take you to a hospital, whether you like it or not.

If I believe you're a threat and you're showing signs that you're a threat, then I can medicate you. I can sedate you. I can, I don't need a cop to tell me to do that.

That's in my rule book. I'm allowed to, to chemically sedate you or restrain you with physical restraints and get you to a hospital.

They've done that up in Canada. And that's what's spooky. If their side wins this battle, that's where I've been seeing it going. Whenever I hear the news saying, "These people are a threat to grandma." "These people are a threat to society." "These people are a threat to themselves."

If they convinced the rest of the world that that's the case, then concentration camps aren't that far away. Like they're just not, they'll look like medical concentration camps, but it'll still be a concentration camp.

So, it concerns me.

**Q:** Do you have backups for all you are documenting? I know they took you off Twitter for a while.

**A:** They've deleted so many accounts, like TikTok. I think I've been through fifty-plus accounts. I can't log onto Facebook anymore. Like they, they don't let me go on Facebook. Google itself doesn't let me use my credentials to get on Google.

Can't even Google. Twitter's the only place that I can actually talk. So . . . that's a main platform.

I wrote my patient care reports, they've all been filed. Now, I'm just repeating the things I've seen and anything I see new—like I've been seeing more what's called like trigeminal neuralgia, the suicide disease. We've been seeing a lot more of that.

I didn't ever run into that before the shots came out. And then I looked it up. It's actually on the Pfizer documents as a thing.

**Q:** What is that?

**A:** Trigeminal neuralgia. A lot of them complain of tooth pain that goes into their head. It's this constant pain that they complain of . . . it's just some of them have an annoyance that's constant. And some of them have excruciating pain that's constant.

It's basically the fifth cranial nerve. And it just causes, it's one of the most severe pains you can ever have. It's horrible. And they call it the suicide disease.

**Q:** Have you seen an increase? Have you seen increased levels of suicides?

**A:** Oh yes. That from the start of the lockdowns and whenever they isolated people, I mean, I've lost, I haven't lost any friends to Covid. All of my friends are frontline workers, nurses, doctors, paramedics, EMTs.

I have lost not one friend to Covid, but I've lost quite a few to suicide since the start of Covid.

**Q:** I'm sorry to hear that, Harry. Can you tell me how you see the overall situation? Is it staying the same, getting better, getting worse, or not sure? Vaccine deaths, vaccine injuries, all of this.

**A:** I mean, it's, in my opinion, I'm giving it to God, but in my opinion, it's going to get worse because we're going to have a lot of these young people that normally wouldn't be experiencing this. I was talking to someone at just a small camp that I was at the other day in Alaska, you know, in their thirties, early thirties, she was thirty-four

I think . . . and was just diagnosed with congestive heart failure at thirty-four years old with no other health history. And whenever I mentioned to her, "Hey, did you take any of the vaccines?" "Yeah, I took two."

She took the Moderna, two of them. And she didn't even know that it could cause heart problems. So, a lot of these people just still don't have the information to tie into their new problems.

The amount of people out there that are running around that normally would be, you know, in their sixties or seventies before ever dealing with these heart problems, those in my opinion, are going to catch up. And there's going to be this like tsunami. There's going to be a big wave. And I think we're all just sort of sitting on a ticking time bomb, just waiting for it to go off. And I'm praying that I'm wrong and for God to get us through it.

Especially when you're dealing with these weird, rapid cancers, you want to call it turbo cancer, but as soon as you do, you're, you get ridiculed for using the word "turbo," but it accurately describes what I've seen. People that have a headache last week or the week before, and then suddenly you're in stage-four cancer and they're young. They're not old. They're just—*immediate* stage-four cancers.

And the people that I know that work in oncology are saying that they're seeing it across the board as well. And a lot of them are quite spooked.

**Jim**: Harry, can you, this is kind of personal, but did this experience, has it brought you closer to God or further away from God?
**A:** Closer. If it wasn't for God, I'd be in trouble.

## Chapter Five

# How They Did It:
# A Roaring Tidal Wave of Money

*Propaganda must be total. The propagandist must utilize all of the technical means at his disposal—the press, radio, TV, movies, posters, meetings, door-to-door canvassing. Modern propaganda must utilize all of these media. There is no propaganda as long as one makes use, in sporadic fashion and at random, of a newspaper article here, a poster or a radio program there, organizes a few meetings and lectures, writes a few slogans on walls: that is not propaganda.*
   —Jacques Ellul, *Propaganda: The Formation of Men's Attitudes*[1]

*If Pfizer had a TV commercial for its Covid vaccine listing adverse events, 158,893 in first twelve weeks . . . the announcer would be reading for 80 consecutive hours.*
   —*Pfizer Documents Analysis Reports*, War Room/Daily Clout[2]

*Only two countries as of 2008 allow direct to consumer advertising (DTCA): the United States and New Zealand. The rhetorical objective of direct-to-consumer advertising is to directly influence the patient-physician dialogue. Many patients will inquire about, or even demand a medication they have seen advertised on television.*
   —Wikipedia, Pharmaceutical Marketing[3]

*The nation's top infectious disease expert and leading scientific voice during the pandemic, Anthony Fauci, M.D., stars in the first phase* *of the government's $250 million campaign to build public confidence in the COVID-19 vaccine.*

*The first $500,000 effort will be followed by a $36.6 million fast-tracked "Slow the Spread" radio campaign set to begin Dec. 21, followed by more radio, print, digital and social media ads the following week.*

—FiercePharma.com[4]

## The Smoking Gun

Most people don't realize how this industrial genocide could have happened, or what made it so ferocious and totalitarian in scope. The part they can't fathom is how all the responsible agencies could have backed the vaccine agenda in lockstep and unison, unless of course, they believed what they were saying.

I can show you how it happened.

In 2021, Maggie and I were racking our minds, attempting to comprehend the overnight transmogrification of my profession—obstetrical care. As an attorney, Maggie has experience in unearthing fallacious documents from major corporations attempting to defraud their customers.

In order to inject pregnant American women with the untested shots (wrongly called "vaccines"), there were three trusted flagship organizations the planners of Covid had to co-opt and capture. They were: The American College of Obstetricians and Gynecologists (ACOG), the American Board of Obstetrics & Gynecology (ABOG), and the Society for Maternal-Fetal Medicine (SMFM).

Incidentally, before Covid, these three flagships never agreed on anything. But, during the Covid debacle, they literally used the identical language, word by word, in what amounted to ideological mantras, repeated ad nauseam, with no struggle, thought, discourse, data, or "science" behind the sterile rhetoric.

Maggie and I knew that these "pillars" of obstetrical medicine—now firehoses of Covid propaganda—had been corrupted and captured, but we needed proof.

We submitted a Freedom of Information Act (FOIA) request for communications between HHS, CDC, and ACOG. Maggie is very skilled and experienced in FOIA requests and knew precisely what to ask for. She had established a trusted relationship with the HHS/CDC individual responsible for negotiating this request. That, I believe, was the key to our success.

Then we waited.

About nine months later, the email we'd been anticipating finally arrived. We were quite astonished to find a staggering 1,400 pages of documents between HHS, CDC, and ACOG. As soon as we begin to study them, we realized about 50 percent were redacted. Still, we knew what we had:

It was the smoking gun.

The FOIA documents revealed that ACOG had signed a "cooperative care agreement" to push the government's Covid agenda and narrative. In order to secure the millions of dollars over the course of several years from HHS and CDC, they were to *never* deviate from the HHS/CDC narrative on COVID-19.

For starters, there was an $11.8 million grant from the Department of Health and Human Services (HHS) to ACOG. But there was much more.

Approximately sixty thousand Ob-Gyn doctors, spanning two continents, are paying members of ACOG. They were to play the role of "trusted messenger" to relentlessly push Covid shots on pregnant women.

And ACOG is not alone—of approximately 275 organizations listed by HHS as COVID-19 Community Corps founding members, twenty-five are health and medical organizations. Eleven other influential "founding member" medical organizations included the American Medical Association, American Nurses Association, American Medical Women Association, and the American Academy of Pediatrics.

## HHS and COVID-19 Community Corps: Follow the "Communication Science"

Essential to its strategy, HHS sought to identify trusted community leaders, enlist them to join its COVID-19 Community Corps, and then utilize these "trusted sources" to convince those around them to take the COVID-19 "vaccines."[5] According to a December 23, 2020 article published by CBS News, HHS ran "focus groups" to fine-tune its pro-vaccine message for what then HHS Deputy Assistant Secretary Mark Weber referred to as "the moveable middle." Secretary Weber also reportedly noted, "Communication science says you need a messenger who resonates as trusted."

Their focus was on finding people with not just local, but also uniquely interpersonal influence. As Harvard public health professor, Jay Winsten, who has advised previous administrations, reportedly explained to CBS News in its December 2020 article, "You want to go for the low-hanging fruit, those that are easiest to pick and harvest." Noting that the focus should be on finding locally influential people to push the vaccines, Winsten added,

"People trust their own doctors, their own nurses, their own pastors, their own social networks. That's very, very different from a distant figure."

The marketing methods utilized by HHS to push the COVID-19 "vaccines"—including the creation of COVID-19 Community Corps—were so vastly different from any other HHS propaganda effort that an article was published in the *Journal of Health Communication* in April of 2022 detailing the process. Featuring now-retired HHS Deputy Assistant Secretary Mark Weber, as lead author, the article confirms that HHS did, in fact, target interpersonal relationships.

As Weber and his coauthors explain:

> Market research impacted every element of the Campaign from the beginning—from overall strategy to early paid advertising, social media postings, and other mass communication strategies. The need for interpersonal interactions with physicians, ministers, family, and community members was clear from the initial market research conducted in the fall of 2020. While the first phase of the Campaign initially focused on mass media messages, it shifted to a more trusted messenger, and community orientation, with outreach focused at the community level.[6]

Weber and his colleagues' "vaccine" marketing efforts were so successful that, after retiring from HHS, Weber apparently formed his own private company aimed at "Achieving bold goals at the Federal Level"—in typical revolving door fashion.

According to Weber and his coauthors, the HHS campaign to push the COVID-19 "vaccines" entered its third phase in 2022 and has evolved into a highly targeted approach using both paid and "earned" media strategies.

The HHS campaign: "Focuses more on precision marketing to identify subgroups with vaccine hesitancy, working directly with communities and using trusted messengers in those communities to deliver messages without the Federal government being directly involved (even though the information may come from a federal source)."[7]

Notably, the article neglects to fully explain—or even recognize—that what HHS is engaged in, is coercive, deceptive, and unethical. This is because HHS used people and methods targeting trust within.

ACOG jumped on board as a founding member of COVID-19 Community Corps[8]—ultimately receiving millions in HHS/CDC grant money[9] and later recklessly endorsing COVID-19 vaccination in pregnancy, even though the clinical trials failed to include pregnant women.[10] *ACOG* began strongly recommending COVID-19 vaccination in pregnancy in July 2021.

## How Women's Trusted Ob-Gyns Became Robots Reading Scripts

ACOG's sixty thousand members steward one of the most trusted and intimate physician-patient relationships in all of medicine, thus providing tremendous opportunity for wielding influence over the "vaccine-hesitant."

What this meant on the ground was that when pregnant women visited their Ob-Gyn, they were subject to preordained scripts that were designed to break down their instincts, which were pathologized as "hesitancy." Additionally, should the patient refuse the Ob-Gyn's recommendation to take the COVID-19 vaccine, the ACOG conversation guide recommended that the clinician document the refusal in the patient's medical record, and during subsequent office visits, push the vaccine again.[11]

COVID-19 "vaccination" Key Recommendations from ACOG's clinician "Conversation Guide" are as follows: *The American College of Obstetricians and Gynecologists (ACOG) strongly recommends that pregnant individuals be vaccinated against COVID-19* (emphasis added). Given the potential for severe illness and death during pregnancy, completion of the initial COVID-19 vaccination series is a priority for this population.

- mRNA COVID-19 vaccines are preferred over the J&J/Janssen COVID-19 vaccine for primary series, additional doses (for immunocompromised persons), and for booster vaccination.
- *ACOG recommends that pregnant and recently pregnant people up to 6 weeks postpartum receive a bivalent mRNA COVID-19 vaccine booster dose following the completion of their last COVID-19 primary vaccine dose or monovalent booster.*
- *Vaccination may occur in any trimester, and emphasis should be on vaccine receipt as soon as possible to maximize maternal and fetal health. This recommendation applies to both primary series and booster vaccination.*
- *For patients who do not receive any COVID-19 vaccine, the discussion should be documented in the patient's medical record. During subsequent office visits, obstetrician–gynecologists should address ongoing questions and concerns and offer vaccination again.*

- COVID-19 vaccines may be administered simultaneously with other vaccines, including within 14 days of receipt of another vaccine. This includes vaccines routinely administered during pregnancy, such as the influenza and Tdap vaccines.
- Pregnant patients who get vaccinated should be encouraged to sign up for the v-safe safety monitoring program of the Centers for Disease Control and Prevention (CDC).[12]

ACOG's income is tracked through ProPublica Nonprofits Explorer.[13] During the seven years from 2011 to 2017 the average annual income from "program services" was stable at about $17 million per year. In 2018, there was an abrupt addition of $28 million to bring the total program services income to $45 million per year, as noted on the graph below. There were also increases in the contribution and royalty incomes during these same time periods.

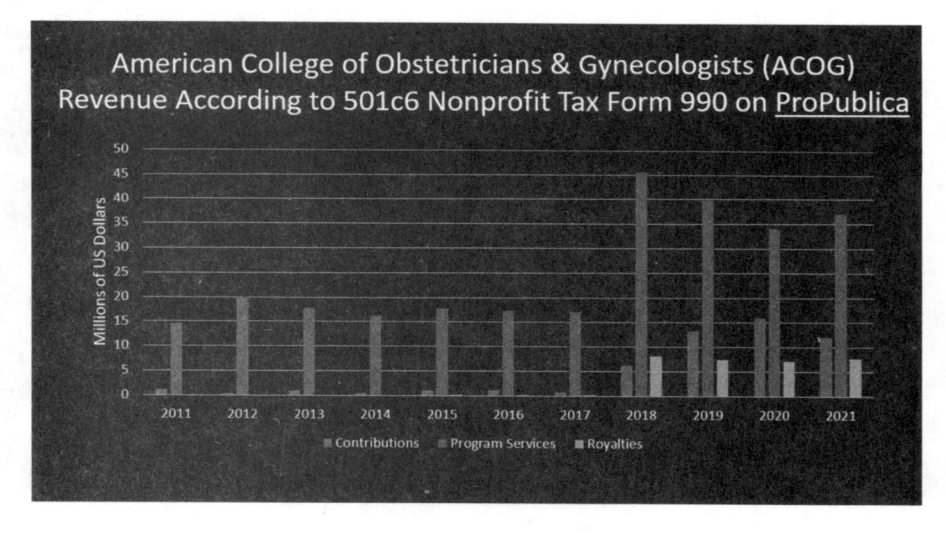

**American College of Obstetricians & Gynecologists (ACOG) Revenue According to 501c6 Nonprofit Tax Form 990 on ProPublica**

How did this abrupt rise of $28 million in 2018 come about? Where did it come from? Is it possible that CDC/HHS was laying the bait for the total capture at that time? Were the plans already in the works for ACOG to push two more additional vaccines in pregnancy—COVID-19 and RSV—in addition to the influenza and Tdap vaccines?

ACOG was the larger of the organizations and founding member of the COVID-19 Community Corps (CCC)—an unprecedented HHS-led, all-encompassing propaganda blitz with a $13 billion budget, that captured every facet of society in the spring of 2021. Named the "We Can Do This" campaign, it was way more advanced and aggressive than any public health campaign in our history.

If you were to click on the link "Media Partners" at wecandothis.hhs.gov website, you will see a list of 3,907 such "partners," and it will take your breath away.

These are a few of the beneficiaries of big government money, for which, it is safe to assume, solid propaganda would be delivered, and was: AARP, Amazon, Buzzfeed, Disney, Facebook, Google, Hulu, Motherly, *The Christian Post, Parade, Parents*, and even Animal Planet! There are hundreds and hundreds of radio stations, religious groups, veteran's groups, and labor unions, to name but a few categories. There is a clear extra bombardment to convince certain groups, such as Blacks, Asians, Hispanics, and Native American groups. "The effort is driven by communication science and provides tailored information for at risk groups," the website boasts.

They were definitely showing a preoccupation with targeting minority groups. Ask yourself: Why? Sixty-one Black radio stations were listed as "partners." In Black neighborhoods, they pulled stunts like hiring retired, former NFL stars to go door to door—sometimes with Anthony Fauci in tow—to ask if everyone in the household had taken their Covid shots yet. If not, they would try and convince them that they should.

This should be illegal.

In addition, there were eighty-three faith-based groups, including Evangelical Christian, Catholic, Muslim, Jewish, and even Greek Orthodox groups. Thirty-nine Asian organizations. Ten Native American. Fourteen veteran groups. Among the labor unions, the Teamsters, United Steel workers, and even the Brotherhood of Railway Signalmen were targeted as well. A long list of "esteemed" medical organizations, of course, starting with the American Academy of Pediatrics, and the American Nurses Association. . . . The NAACP, the National Bar Association, and a dubious outfit called "Nurses for Biden/Harris . . ." The list is a Who's Who? and a Who Else? of the entire fabric of American professional society.

How much money did they receive?

It's impossible to separate it out, but one can look at the whole and sometimes identify specific tentacles.

According to the *New York Times* article of March 11, 2022, there was $5 trillion in pandemic spending without knowledge of where it went.[14] Over $186 billion was given to the health-care sector alone, with more than 420,000 payments made to health-care systems, individual hospitals, and other related entities.[15,16]

## Another Tragedy—Teenager Killed by the Shot

Allen Martin, whose teenage daughter Trista was killed by a Pfizer shot that she got in order to attend a concert, is featured along with his grieving wife, Taylor, in the 2023 documentary, *Shot Dead*. The film highlights stories of parents whose children were killed from the injection. On Martin's twitter feed, he has uncovered some instances of direct payments that so-called influencers received, to fight "Covid disinformation" on the internet—to bully the victims and promote the lies. Allen and Taylor Martin and many other parents including Earnest Ramirez, Dan Hartman, and countless others whose children were killed by the Covid "vaccines," are frequently mocked, derided, and called liars on their social media posts. The cruelty and insensitivity of these vicious attacks on these suffering parents by the nefarious, pharmaceutical-funded operatives is beyond evil—it is unfathomable. Only the most vile sociopath could engage in these types of attacks.

In the Covid era, sadism was quickly normalized in the mass culture and on social media.

A Canadian Law Professor, Tim Caulfield, from the University of Calgary, was often interviewed on Canadian mass media about Covid. Martin's tweet, superimposed over a photo of the professor, reported some jaw-dropping numbers paid to Caulfield:

- In April 2020, he received $381,000 in federal and provincial grants to combat Covid "misinformation."[17]
- April 2021, he received $1.75 Million from Federal Health Minister to counter vaccine "misinformation."[18,19]
- According to the Government of Canada Grants and Contributions website, Caulfield received $3 million between March 2, 2022 and July 1, 2023.[20]

## My Oklahoma Senate Testimony

Those who are killed and injured by the vaccine are my primary focus. They're the source of my horrible emotional (and physical) pain—my raison d'être in these final years of my career. My life, really.

I was asked to testify at the Oklahoma Senate meeting on March 26, 2024, along with other experts and parents Allen and Taylor Martin, mentioned earlier, who had lost their daughter, Trista, to the vaccine. Despite my disability and physical pain when traveling, I was committed to being there in person.

When Allen and Taylor picked me up at the airport, we hugged, cried, and prayed together. It would have been Trista's twentieth birthday the day before, Monday, March 24.

We headed to the Capitol Senate building and testified. Each testimony was limited to five minutes. I had much righteous anger and emotional pain—it felt almost as if Trista was one of my daughters, even though I had never met her. My testimony was passionate, to say the least. I did not remain seated as did all the other experts.

I stood up and screamed out the statistics, wearing my T-shirt honoring Trista—the same one her mom and dad were wearing. I was speaking on behalf of the Martin Family, who were representing all the other Oklahomans that have been vaccine-killed (8,500) and injured (284,000). I blasted the "Butchers: Bourla & Bancel," the HHS, FDA, ACOG, ABOG, SMFM and more, accusing them of collusion, fraud, racketeering, assault, battery, and mass murder, and I called for criminal indictments.

Dr. Peter McCullough texted me shortly thereafter: "Admiral, you took no prisoners, I could hear you all the way down here in Dallas." My vaccine injured friend, Lyndsey RN, communicated a similar message, "I could hear you all the way to the East Coast!"

## An Obscene Amount of Taxpayer Money Spent

A visit to www.USA.spending.gov reveals a post-2020 United States that is essentially a bloated whale of nothing but "Covid spending,"—that doesn't even attempt to quantify what all this money was for. We know what it was for: They bought off an entire society, an entire civilization, to beat the drums about all things Covid—until the propaganda drone overtook people's minds, and if they had doubts, they kept those to themselves. At a time of record inflation and unemployment, Covid propaganda was, for many, the only way to find employment.

It was subsequently reported by many major news outlets including the *New York Times* and the *Washington Post* that $5 trillion US taxpayer dollars were spent since the Covid pandemic started, and these news outlets were unable to account for how and where it was spent. A large proportion of these monies were likely spent on "quid pro quo" under-the-table illegal contracts with medical organizations, non-governmental organizations, large healthcare systems, faith leaders, legacy media, sports associations, medical journals, local and national influencers, minority leaders, unions, sports leagues like Major League Baseball, National Football League, NASCAR, NAACP, GLAAD, and many others.

Maggie and I have published about fifteen articles on the *America Out Loud* platform over the past eighteen months exposing the governmental and institutional corruption; these are easy to read, highly cited medical-legal briefs. Of note was our article on May 7, 2023, documenting the governmental capture of the American College of Obstetricians and Gynecologists (ACOG)—a "founding member" of the COVID-19 Community Corps, to promote the safety and efficacy of the vaccine. ACOG has jurisdiction over sixty thousand Ob-Gyn physicians in US, Canada, and several countries in South America. Within a few months after ACOG was captured, the American Board of Obstetrics & Gynecology (ABOG) and the Society for Maternal-Fetal Medicine (SMFM) were also captured and colluded to threaten sixty thousand Ob-Gyns with harsh punitive measures including loss of board certifications and loss of state licenses if they spread "misinformation"—implying that if anything other than their insinuation that the "vaccines" were safe, effective, and necessary, even in pregnancy, the physicians would be punished.

ACOG/ABOG/SMFM, perhaps more appropriately renamed the "Global Mother-Baby Death Cult," continue to this very day to push this lethal, injurious drug to the most vulnerable population—pregnant women, preborns, and newborns. Ob-Gyn doctors should know better, especially after the thalidomide and DES disasters of the 1950–1970s—the age-old "Golden Rule of Pregnancy" was **NEVER** to be violated; that is, never, ever use novel substances in pregnancy.

Very few Ob-Gyns have stood up to this violation of medical ethics—they continue collecting their paychecks, remaining silent, and violating their Hippocratic Oaths. They also know from the multiple past publications in our specialty that anything that causes inflammation in pregnancy is disastrous; COVID-19 "vaccines" were known to be one of the most inflammatory substances ever injected into the human body.

## Capture of Hospitals, Health-Care Providers, and Faith Leaders

Maggie and I published another article on December 10, 2023, documenting the governmental capture of more than 420,000 health-care systems, hospitals, and health-care providers across the US during the pandemic.[21] About $186.5 billion was allocated to various programs to accomplish their capture and recipients likely engaged in quid pro quo arrangements like that of ACOG. The State of Texas was targeted with Houston Methodist Systems accepting about $350 million, the first hospital system in the country to mandate vaccinations in their employees. The rest of the nation's hospitals fell like dominoes to this forced employee "vaccination tyranny." As detailed elsewhere, I was fired from SSM Health System of St. Louis, MO a nearly $10 billion-dollar Catholic Hospital System—they accepted $306.9 million in early 2021 and SSM Health followed suit with the tyrannical measures of forced employee vaccination. I was given a religious exemption.

Even more disturbing was our article published January 14, 2024, detailing the evil capture of a large number of major faith leaders of all traditions, orchestrated by White House Chief of Staff, Jeff Zients, then Director of the NIH Frances Collins, and Surgeon General Vivek Murthy on a 3.5 hour symposium in May of 2021 entitled "Faiths4Vaccines."[22] They made blatantly false scientific statements to these faith leaders to force them to push vaccines in their congregants and eliminate vaccine hesitancy. There were many contorted spiritual references that were made in this symposium, including that God's "miracle cure" for the pandemic was the COVID-19 vaccines. Multiple Biblical scriptures were grotesquely twisted; perhaps the most stunning to me was the verse from Deuteronomy 26:8 below:

> *The Lord took us out of the narrow place with a strong hand and an outstretched arm.*

The government's *Faiths4Vaccines* "Toolkit" interpreted this verse for the faith leaders in this national symposium, sponsored by White House Chief of Staff Jeff Zients:

> "The narrow Place" refers to Covid.
> "A strong hand" refers to the hand holding the syringe.
> "An outstretched arm" is the arm receiving the vaccine.
> The Bible's message is clear: Go forth and be vaccinated, that ye may live!

In January 2024, Maggie and I published an article on *America Out Loud*[23] demonstrating that powerful US government officials, including Chief of Staff Jeff Zients, Surgeon General Vivek Murthy, and NIH Director Francis Collins used religion in an attempt to convince faith leaders around the country to push the COVID-19 injections. A deep dive into the organization *Faiths4Vaccines*—a founding member of the HHS's vaccine-propaganda machine COVID-19 Community Corps—has revealed just that.

That the HHS was tapping faith leaders in the spring of 2021 to push the uptake of COVID-19 vaccines was not a surprise. We uncovered this in a previous article that we broke on the COVID-19 Community Corps at the end of 2022.[24] But what did surprise us as we dug deeper for this article was the extent to which faith leaders were pursued to push the COVID-19 vaccines and the inappropriate—if not unconstitutional—manner in which government officials persuaded them to push the shots.

With eighty-six founding members, the "faith leaders" category of the COVID-19 Community Corps was the most numerous. These founding members included both individual faith leaders and faith organizations from a variety of religions—including the American Baptist Church, Catholic Charities USA, the Episcopal Church, the National Association of Evangelicals, the Greek Orthodox Archdiocese of America, and the New York Jewish Agenda, just to name a few.

Not surprisingly, many faith organizations received federal money during the pandemic. For example, an entity called "American Baptist Churches in the USA" reportedly received $1.5 million in COVID-19 relief bailout money—in the form of two forgivable "loans" that spanned 2020 and 2021.[25] In their defense, the country was shut down for much of 2020, which left faith organizations, some of whom rely on member donations to pay day-to-day costs, facing financial disaster.

In our opinion, however, it wasn't financial incentives that ultimately convinced faith leaders to push the COVID-19 vaccines on their members. Instead, it was a strategy that baselessly appealed to faith leaders' deeply held moral and religious beliefs—targeting their virtues instead of their vices—and which shamelessly relied on religious doctrine and a pro-vaccine "theological" interpretation to support pushing the shots.

For example, at a May 2021 national summit for faith leaders, NIH Director Francis Collins—referred to as "Reverend-Doctor"—would address hundreds of faith leaders across the nation, claiming that the COVID-19 vaccines were God's literal "answer to prayer" and urging faith leaders to believe that pushing the shots was a "love your neighbor moment." In what was a sermon-like address tailored to appeal to them, Collins admonished them not to believe "conspiracy theories" about "possible side effects"—which Collins falsely said were untrue.[26]

In the spring of 2021, when COVID-19 vaccine uptake had leveled off, faith leaders seemed to be the Biden administration's answer to getting hesitant Americans vaccinated. The reason? Faith leaders had vast untapped potential to convince vaccine-hesitant Americans to take the shots—particularly if they could be persuaded that COVID-19 vaccination was a moral obligation owed to others. Like health-care providers, faith leaders were "trusted"—but arguably much more so than one's doctor. Faith leaders were associated with the divine. They were often connected to deeply personal, intimate, and even sacred moments in the lives of Americans—spiritual moments that involved life and death, great joy, and deep sorrow—such as baptisms, christenings, bar and bat mitzvahs, weddings, and funerals. As noted in one peer-reviewed article about the role faith leaders played in creating vaccine confidence, faith organizations were able to penetrate people's vaccine hesitation in "hyperlocal" ways. Moreover, religion in the United States is widespread—Wikipedia reports that "An overwhelming majority of Americans believe in a higher power, engage in spiritual practices, and consider themselves religious or spiritual."[27] Consider that the American Baptist Churches USA alone claims 1.3 million members and approximately five thousand congregations, with 42 million Baptists globally.

"Looking back, it appears that governments began to tap faith leaders to push COVID-19 vaccinations as early as February 2021. That month, Washington, DC's Mayor Muriel Bowser announced a pilot initiative that was disturbingly called '*Faith in the Vaccine*'—a program in which DC Health would partner with Washington's faith communities to try to get people vaccinated."[28] The government's initiative to partner with faith organizations would eventually be pursued on a national level.

*Faiths4Vaccines* describes itself as a "multi-faith group of local and national religious leaders" seeking "to increase opportunities for faith-based institutions, particularly houses of worship, to engage and support the United States government in its efforts to increase vaccination rates" and combat "vaccine hesitancy." *Faiths4Vaccines*' top shared goal is shocking—it strives to "Demonstrate 'religious communities' trust in the vaccine."

According to a peer-reviewed study that assessed the impact of faith organizations on COVID-19 vaccination uptake, *Faiths4Vaccines* includes over one thousand faith leaders across the United States. This study found that *Faiths4Vaccines* held thirteen "bi-weekly roundtables" led by faith leaders who used their houses of worship as vaccination sites and had regular engagement with the White House, including the White House COVID-19 Task Force, as well as the CDC and HHS, as part of these bi-weekly calls. A spin-off initiative of *Faiths4Vaccines*—called *Youth4Vaccines*—pushed the COVID-19 vaccine among America's youth and also held a roundtable "showcasing how youths of faith are leading in their communities within the COVID-19 vaccination efforts."[29]

The COVID-19 vaccines, however, have hardly been God's answer to prayer for the estimated 585 million global citizens around the world who have been injured or killed after taking them. These include Maddie de Garay, who was only twelve years old when she suffered extreme neurological injuries following her second Pfizer COVID-19 injection while she was participating in the clinical trial. For Maddie and her family, and countless others like them, the vaccines have been a horror story, not an answer to prayer. Although Maddie's injuries left her in a wheelchair, they were reportedly downplayed by Pfizer as only "stomach issues"—casting grave doubt on the rigorousness of the clinical trial standards conducted by Pfizer.[30]

Maddie is not alone. As the many faces and stories of vaccine-injury websites like *RealNotRare* demonstrate, there are countless stories of COVID-19 vaccine-injured individuals who have been ignored, gaslighted, and discarded by their government. For them, the vaccines have been a living hell that no one should have to experience.

When government officials and those who pushed the shots in cooperation with the government fail to acknowledge the shots' spectacular failure, and when they refuse to give voice to or help the millions who have been hurt and damaged by the shots they pushed—this too shocks the conscience. In manipulating faith leaders to push the COVID-19 injections on their fearful members, our government has abused God, abused Scripture, and abused religious doctrine.

With an unlimited supply of "We the People's" tax dollars, almost any lie can be promoted. The overreach by our government at all levels up to the White House, their gross manipulation and bribery of all sectors of our society, was incomprehensible. It is rather astounding that 25 percent of our population saw through the lies from the beginning and never took a vaccine. They withstood the pressure—the pressure of a propaganda campaign that infused the entertainment world as well.

## Propaganda on Steroids

Late-night talk host and vaccine enthusiast, Stephen Colbert, went as far as having full-body syringe suits made—and transformed his whole program into a creepy, Broadway-like musical number. With Colbert as "Head Needle"—he was accompanied by a whole dance troupe—wearing silver syringe leotards—who weaved themselves into the audience singing the praises of vaccines. Disgusting.

Our Centers for Disease Control (CDC) created the social media status stickers boasting, "I Got My Covid Vaccine," a sinister psyop if ever there was one.

Here are some excerpts from www.wecandothis.HHS.gov website:

> The HHS COVID-19 Public Education Campaign is a national initiative to increase public confidence in and uptake of COVID-19 vaccines and educate

the public about the availability of COVID treatments while reinforcing basic prevention measures.

Through a nationwide network of trusted messengers and consistent, fact-based public health messaging, the Campaign helps the public make informed decisions about their health and COVID-19, including steps to protect themselves and their communities.

The effort is driven by communication science and provides tailored information for at-risk groups.

**Strategy and Goals**

This effort focuses on Americans who want to protect their health, and may have questions about the COVID-19 vaccines. We aim to:
- Explain how Americans can protect themselves from COVID-19.
- Strengthen public confidence in the vaccines so people have the science-based information they need to make decisions about getting vaccinated and get protection from the worst outcomes of COVID-19.
- Increase vaccine uptake by informing Americans about how and where to get vaccinated.[31]

**COVID-19 Community Corps**

The campaign is expanding its reach by engaging with a broad range of groups and individuals, including trusted community organizations, local leaders, and others sharing the goal of increasing vaccine confidence and uptake.[32]

**Public Education Activities**

Our public education activities are organized around these themes:
- **Building Vaccine Confidence**: Information and resources to build confidence about the COVID-19 vaccines so people are ready to get vaccinated when it is time.
- **Protecting the Nation**: Fact based answers to common questions about vaccine development, safety, and effectiveness.

This is easily the most appalling propaganda campaign in our nation's history, and it hit our country like a tsunami, seemingly out of nowhere. It's no wonder that pregnant women, in wanting to protect their own health and that of their unborn child, would fall prey to this coercion.

The endless commercials, social media ads, posters, flyers—one could look at almost no surface of American public life after April 1, 2021, and *not* see COVID-19 vaccine propaganda. The panopticon made sure we displaced all other thoughts and thought only of *it*.

Covid was the revolution—the future, the everything. Nothing existed before it.

That's how they did it.

In the manner of a revolution, something new that could not be questioned. Nothing, but nothing, would get in its way, or thwart its ferocious will.

The Covid revolution displaced, seemingly overnight, all other values, crushed them for parts, and rose up with one singular human aspiration and goal: *get vaccinated.*

Against this $13 billion propaganda blitz, the competent, trusted, well-liked physician with no blemish on his record stood no chance. By questioning anything, we became a threat to the status quo. I was reduced to an outcast in my own practice. One of the nurses felt so threatened by my "vaccine hesitancy" that she made herself a victim, and persuaded many staffers that she was in some way being tormented by me and my stance. They would huddle around her and cast angry glances my way, making it impossible for me to say anything, without standing accused of harassing her, just by my very being. The entire hierarchy had been flipped on its head. I knew I was powerless but kept myself focused on never returning any of her bizarre aggression. I could see these manifestations of Covid hysteria were creating an atmosphere of tremendous threat for anybody who did not adhere to the diktats of the Covid revolution.

## "Communication Science" Means Access and Influence

Regarding the issue of "trust," a pregnant patient's relationship with her Ob-Gyn is arguably one of the most intimate and sacred physician-patient relationships in all of medicine. This is not without reason—as one patient and writer notes: "They're right next to you for the most momentous occasion of your life."[33] Pregnant mothers trust their Ob-Gyn doctor with the most intimate and sensitive information about their own bodies, their sex lives, and, if pregnant, about the new life growing inside of them. Their Ob-Gyn is one of the first persons to actually see a mother's newborn baby, whether reading prenatal images during the pregnancy or during the birthing and delivery process. Perhaps not surprising, some individuals have even confessed the development of a non-romantic affection for their Ob-Gyn that rivals that of the baby's father in some ways, due to the "complete vulnerability" many women reportedly experience with their gynecological and pregnancy specialists.[34] In sum, government capture of ACOG would provide access to and influence over near-perfect "trojan horses" to market the CDC's pro-vaccine message.

As Ob-Gyn patients, women have been referred to as "A Brand's Powerhouse."[35] This is not without good reason: Marketing studies have shown that women reportedly make a full 90 percent of all health-care decisions within their household.[36] Convincing women to take the COVID-19 shots was almost a guarantee that they would become pro-COVID-19 "vaccine" messengers within their own families.

Moreover, if the COVID-19 "vaccines" were considered safe enough to administer to pregnant patients (and thereby trans-placentally to their unborn babies)—certainly they were safe enough for everyone. If HHS and CDC could pull off government capture of ACOG, and convince its Ob-Gyn members to push the shots on their patients, this would be a bonanza for reaching the "vaccine" hesitant—what HHS Deputy Assistant Secretary Mark Weber referred to as the "moveable middle."[37]

## But, Why?

Let's ask the central question: Why was it so crucial that the US Government went to such lengths, spared no expense, stopped at nothing, literally, to get these injections into its citizens, and particularly those who would be most harmed by it—unborn babies?

The answers, when we begin to trace them, are extremely unsettling.

Referring to her stellar team of volunteer researchers at *The Daily Clout*, Dr. Naomi Wolf said, in an interview with Jason James, at *Brave New Normal*: "What they found in the documents is the greatest crime against humanity in recorded history . . . with a special emphasis on sterilizing women and destroying the next generation reproductively."[38]

"What's so creepy about reading the Pfizer documents," she went on, ". . . Covid is ostensibly a respiratory virus. There's pretty much nothing studying mucous membranes, or lungs. It's all about sperm and ovaries and sex organs. They literally mated unvaccinated female and vaccinated male rats together, sacrificed the rats and then examined the cells of their sex organs. . . .

"That's how creepily focused it is on reproduction."

Unfortunately, we've found ourselves here before. One can't look back over the past century and see smooth waters for pregnant women. But this time, there seems to be no stopping it, and most terrifyingly, there's neither alarm nor remorse on the part of the responsible institutions.

It has been stated by many since the catastrophe of thalidomide, that it killed or injured 100 percent of preborns when administered at the most vulnerable time in pregnancy. Why have our regulatory bodies and sixty thousand Ob-Gyn physicians, let alone every other physician in the world, ignored the lessons we (should) have learned from the thalidomide disaster?

And even though thalidomide caused far less morbidity and mortality in pregnancy than did DES, the graphic nature of the severe birth defects caused by thalidomide—and the photos documenting those birth defects—are permanently etched in the public's psyche.

## Chapter Six

# Twin Towers of Catastrophe in Maternal-Fetal Medicine: Thalidomide and DES

*Every doctor, every hospital, every nurse has been notified. Every woman in this country must be aware that it's most important that they check their medicine cabinet—that they do not take this drug.*
 —President John F. Kennedy, August 1, 1962, news conference

*Damage varied from day to day. If the mother took the drug around day 20, you'd be getting central brain damage. Day 21, it would be the eyes. [Day] 22—23 it would be the ears, the hearing, and the face. A tablet on day 24 was capable of removing a complete pair of arms, and over the next four days, corresponding levels of leg damage. . . . Thalidomide attacks the embryo almost with the precision of a sniper's rifle.*
 —Dr. Martin Johnson, chairman, the Thalidomide Trust

*The original catastrophe maimed 20,000 babies and killed 80,000: War apart, it remains the greatest manmade global disaster.*
 —Harold Evans, 2014 *Guardian* article, "Thalidomide: How Men Who Blighted Lives of Thousands Evaded Justice"

## "Astonishingly Safe"—The True, Shocking Story of Thalidomide

The thalidomide tragedy of the late 1950s and early '60s serves the function of chemical scapegoat for an "industry" that would like you to believe this was a unique, black-swan event—like the Titanic of modern medicine. It is used to bolster the idea that because of it, all kinds of rules, regulations, and safety tests were put in place to pave the way for the wraparound Pharma-Nanny State we live in today.

The truth is the inverse, like the negative of a photograph: Thalidomide was actually a high-water mark in a pharmaceutical regulatory climate that not soon after, gave way to decadent ruin. When we take a close look at all the moving parts, players, and features of the international drug approval and marketing worlds that collided then, we see responses and standards we can no longer expect in the post-Covid, free-fall world.

Thalidomide happened at a time in our history when the chemical and pharmaceutical "industries" did not yet own and control the FDA, the courts, and the global media, as they do today.

There was, then, such a thing as being horrified and enraged by the actions of a reckless and greedy pharmaceutical company, and by contrast, there was virtually no such thing as *not* siding with the victims. All of that has now been turned on its head.

The thalidomide saga is, in fact, the dark prequel to Covid shots being given to pregnant women, even though the former was a drug, and the latter is classified as an mRNA vaccine.

Few are aware of the true, shocking story behind thalidomide. It opens a formerly closed portal into the history of not just one catastrophe of "medicine" but what it came out of—those intersecting dark chords of history, ideology, international conflict, trade, pharmacology, and post–World War II mass-psychological trances.

What most of us do not know about thalidomide, despite its immense, dark, international fame, is an essential part of how and why we have landed where we are, almost four years into the Covid catastrophe. Except the word "catastrophe" here implies accident, things going wrong. With Covid, one can't see what is up or what is down. One can't ferret out any shared values, of the "old world"—no sense of what is even considered unacceptable, except one single thing: being critical of vaccines. That is unacceptable. Including when vaccines kill within moments of injection.

And in the typhoon of propaganda and gaslighting that began in 2020, leading to the mass injecting of pregnant women in 2021, with the incantation "safe and effective," the damage and death toll would quickly surpass that of thalidomide, widely known as the "greatest manmade global disaster" after World War II.

In 1956, thalidomide was branded as a "wonder drug"—combining the best chemical aspects of sedatives, painkillers, and mood-lifters—and launched upon the world. The harsh reality, however, was that victims of thalidomide spent their lives, as did their parents, fighting for compensation for the chemical crimes committed against them—a drug released around the world that was never demonstrated to be safe for pregnant women, and discovered to be causing extreme birth defects, notably missing arms and legs, as well as death in the womb, at birth, and within a year of birth.

## Greatest Man-made Global Disaster—Silence as a Sign of Shock?

As recently as 2014, one of the great heroes of the thalidomide scandal, British editor Harold Evans, wrote an article in Britain's (very pro-pharma) newspaper, *The Guardian*. It had the title: "Thalidomide: how men who blighted lives of thousands evaded justice," with a heart-wrenching photo of an armless blond toddler in a dress, playing with blocks with her feet. Evans calls thalidomide, rightly, (in 2014), ". . . the worst and biggest crime in postwar Germany."[1] The real numbers are hard to come by but estimated to be at least twenty thousand maimed and eighty thousand killed, either in-utero, shortly after a natural birth, by infanticide, or within a year of birth.

The German company that made the drug—Chemie Grünenthal—apologized formally to the victims in 2012, fifty years after it was withdrawn from the market. The apology came from the company's CEO, Harald Stock, at a ceremony in Stolberg, Germany, where the company is based, and where a bronze statue of an armless thalidomide child was also unveiled:

Stock said:

> We ask for forgiveness that for nearly 50 years we didn't find a way of reaching out to you from human being to human being. We ask that you regard our long silence as a sign of the shock that your fate caused in us. We wish that the thalidomide tragedy had never happened. We see both the physical hardship and the emotional stress that the affected, their families and particularly their mothers, had to suffer because of thalidomide and still have to endure day by day.[2]

When I call thalidomide a "scapegoat," I mean specifically this: You (according to mass media and mass culture) may call this case a calamity, a scandal, and lament the lack of criminal punishment to the guilty parties. But why only thalidomide? Because it appears that the standard does not apply elsewhere, and certainly does not apply to present day vaccine-induced damage, disfigurement, or death.

There was, actually, a criminal trial—a kind of second Nuremberg for those who created and marketed the drug. It began its proceedings in the town of Alsdorf, Germany, in 1968, and had been the reason countless families had to forestall their hopes of compensation—as the criminal trial was the main event.

Grünenthal's excuses, deflections, denials, and evasions were astonishing, and included: that the deformed babies were "acts of god," that they were caused by the mothers drinking alcohol, and that they resulted from botched home abortions. The company insisted for a long time that it was blameless. "It had the discreet support of the politically well-connected chemical industry, mindful that a conviction would raise insurance premiums," wrote Harold Evans in *The Guardian*.[3] "The North Rhine-Westphalia public prosecutors found the company obstructive. They had to seize the most important Grünenthal documents in police raids on its "bunker" and a company lawyer's house."

Evans reported that it took six years to examine five thousand case histories, from mothers who had taken the drug while pregnant and given birth, to deformed or dead babies, as well as men and women who had suffered an additional thalidomide effect—irreversible nerve damage. "The bill of indictment they prepared against nine Grünenthal employees ran to 972 pages," Evans wrote. "In support, they had lined up 351 witnesses, 29 technical experts and 70,000 pages of evidence. There were 400 co-plaintiffs."

Grünenthal came with an army of its own lawyers—forty in total—outnumbering the prosecution's. There were so many people in attendance, including hundreds of spectators, the trial had to be held in a casino.

Two and a half years into the trial, something inconceivable happened: On December 18, 1970, the whole trial was shut down by the German government, and the nine men charged with "intent to commit bodily harm" and "involuntary manslaughter" walked free. Evans reported that, "The judges said this was with the explicit approval of the prosecution. They granted Grünenthal immunity from any further criminal proceedings. Silence was imposed on the 2,554 German families who had children with foreshortened limbs or no limbs; many had damaged organs, and some were blind. Their parents then had no alternative but to go along with a miserable compensation scheme contrived by the government and the company. It benefited Grünenthal far more than it did the victims."

Grünenthal eventually announced it would contribute $27 million to the 2,554 German families who would in return give up their civil suits. Evans details in his long article how documents unearthed later revealed how the backroom deals went down, and the fact that acquittals had been judged "unlikely." The state prosecutor for the victims said that he was provided with two young prosecutors as assistants, very late in the game, and ". . . they went behind my back and betrayed the victims."

## History of Thalidomide

It is said that the German family—the Wirtz family—that founded Chemie Grüenthal was in the business of making soaps, perfumes, and cleaning fluids before World War II.

In 1946, right after the war, Hermann Wirtz, described as a devout Catholic, and self-admitted member of the Nazi Party before the end of the war, changed direction, founding the company he called Chemie Grünenthal, with an influx of a new brain trust—friends he'd made during the war, one presumes. He put former Nazi, Dr. Heinrich Mückter, in charge, and he, in turn, hired about a half dozen scientists and technicians. To the last man, these scientists had performed experiments on people in Hitler's labor camps.[4] Otto Ambros, who'd invented Sarin nerve gas and had overseen the construction of the Auschwitz IG Farben plant, became the company's non-executive chairman. He'd been sentenced to eight years of prison at the Nuremberg trials, but, according to Evans, was released after four, ". . . to help the US Army chemical corps."[5]

Evans calls the presence of high-level former Nazis the "distinctive feature" of this post-war pharmaceutical company.

## Miraculously, the US Was Spared

If people know one thing about thalidomide, they probably know that this toxic drug was never licensed by the FDA in the US, thanks to the miraculous intervention of an FDA safety officer named Frances Kelsey.

In the US, the then relatively young agency known as the Food and Drug Administration (FDA) hired a Canadian pharmacologist—Dr. Frances Oldham Kelsey. Kelsey put up the only protection against this wicked chemical in the western world, by blocking its approval at the FDA, for lack of safety data in pregnant women. Although this was her first "assignment," she and her team blocked thalidomide not once, but six times to the immense frustration of the licensing company Richardson-Merrell, which had assumed it would fly right through, as it had done all over Europe and Australia.

One of the things that had troubled Dr. Kelsey was how close in tone to advertising the claims made for thalidomide seemed. "The claims made in the NDA for thalidomide were too glowing," Kelsey wrote in her memoir. "The claims were just not supported by the type of clinical studies that had been submitted in the application." Kelsey and her team stood their ground. *The Independent* UK quoted, in 2015, FDA historian John P. Swann: "It's fascinating to see how many letters and communications went back and forth. She was asking for additional data, and what they were sending her was, in her eyes, insufficient to answer the questions she was raising."[6]

Thalidomide was taken off the market ten days after news broke in a German newspaper, in 1961, that it had been linked to severe deformities. By the following year, 1962, Frances Kelsey had become—rightly—a folk hero, following a front-page story the *Washington Post*, about her courageous act of insisting that a drug company must prove its drug's safety profile. President John F. Kennedy awarded Kelsey and two of her colleagues who worked closely with her—Oyama Jiro and Lee Geismar—the President's Award for Distinguished Federal Civilian Service, in 1962 at a White House Ceremony.

The heroics of safety testing, of increased regulations and safeguards against another potential disaster like thalidomide, the US government taking pride in protecting its citizens—these were the notions that were in the air in 1962 in the nation's capital. In the wake of the thalidomide near-miss, and Kelsey's award, new legislation was also passed that was designed to protect the American people going forward: The US Kefauver-Harris Amendment, or "Drug Efficacy Amendment," was a 1962 amendment to the Federal Food, Drug and Cosmetic Act, that was intended, in the words of the FDA, ". . . that consumers will not be the victims of unsafe and ineffective medications."[7] Signed into law by President Kennedy on Oct 10, 1962, the amendments ". . . established a framework that required drug manufacturers to prove scientifically that a medication was not only safe, but effective."[8]

The amendments, passed unanimously by both houses of Congress, brought more bad news for drug companies—at least in theory: they would now have to report all side effects to the FDA, study subjects would have to give their informed consent, and the FDA would have 180 days (as opposed to sixty, as before) to approve a new drug application. The FDA would also have to retroactively evaluate the effectiveness of all drugs that had been approved between 1938 and 1962. In addition, it transferred to the FDA control of prescription drug advertising, which would have to include ". . . accurate information about side effects."

FDA historian, John P. Swann, considers this the moment when ". . . members of the public really started taking notice and having expectations that the government will protect them. And FDA rose to that challenge fifty years ago."[9]

Did it, though? To be precise, FDA had tossed the thalidomide file to Kelsey, expecting her to rubber-stamp it. Had she been just about anybody else, most likely the FDA would not have "risen to this challenge." And there, but for the grace of God, an untold number of American babies were spared the fate of the damaged or dead babies in other countries around the world. American babies were not entirely spared, however, for a shocking reason: The distributor of thalidomide had handed out some 2.5 million tablets to more than twelve thousand doctors for purported tests while it awaited FDA approval. Some of these doctors simply gave it to their pregnant patients. It is estimated that there are perhaps around one hundred thalidomide cases in the US, but they all had to fight tooth and nail to understand what had happened to them, since officially, the US had been spared the curse. President Kennedy had held an emergency news briefing alerting women to "check their medicine cabinets" and if they saw thalidomide, to dispose of it immediately.

President Kennedy's stark warning that an evil chemical was loose in the land and could be lurking, unsuspectingly, in anyone's medicine cabinet happened at a time when a president could override the agendas of pharmaceutical companies and their subsidiary profiteers.

The thalidomide tragedy is a foreshadowing event which at first glance appears to reveal the moral lesson that back in the era of the late 1950s and early 1960s, "we" did not know that chemicals ingested by mothers could cross the placenta and affect the developing fetus. This was one of the core arguments of the culprit—Chemie Grünenthal, and its lawyers. It was untrue. Instead, the drug had been tested on animals that tragically were the wrong animals, primarily mice. It was said, and it was true, that ". . . you could not kill a mouse with thalidomide." No dose was lethal to a mouse. Many years later, however, the teratogenic effects would be found instead in animals like zebra fish and chicken embryos.

The public outcry, the criminal trial (even though it was thwarted), and the shaming of Chemie Grünenthal were correct, rational, and right.

President Kennedy's actions and the "Drug Efficacy Amendment" were also rational, and right. The FDA basked in its glory, and Kelsey's intervention refracted glory on the FDA it may not have entirely deserved.

Certainly, no presidential medals of courage would have been awarded, in 2020 or 2021, for stopping a dangerous drug or vaccine from harming Americans.

In 2022, when President Joe Biden stood at a podium and barked: "Take your shots!" at the American people, the very backdrop was vaccine advertising: "Vaccines.Gov."

The vaccines.gov website is bereft of any hint of caution about vaccines. The only imperative is to "find a location near you," and if you have any problem with your insurance, there are programs that will pay for it. The bottom of the site carries standard boilerplate text that speaks in the "voice" of a cult:

> "Vaccines can help you stay healthy. Get yours now."
> "How do I schedule an appointment to get vaccinated?"
> "How do I confirm the pharmacy has vaccines in stock?"
> "Learn about how to protect yourself and others."
>
> We want to make COVID-19 vaccination easy and accessible to everyone. We work with partners such as clinics, pharmacies, and health departments, to provide accurate and up-to-date information about vaccination services in your area.

At Covid.gov (an address where the "US government" and "Covid-19" seem to merge into one another), World War II's Rosie the Riveter, the very embodiment of patriotic duty, is retro-fitted for the cause: "**We can do this**."

And so, they did.

From the site, the exhausting propaganda assails the reader—here is but one sample, where the apparatus self-describes its agenda to steamroll the American people:

> **Overall Campaign Strategy**
> HHS is using a comprehensive media campaign and collaboration with public health influencers to reach Americans through traditional, digital, and social media in culturally appropriate ways. HHS is also providing rapidly evolving messages based on the latest scientific information and research.

Messages are specially designed to reach diverse populations most likely to take action to protect their health and those disproportionately affected by COVID-19.

A robust and continuous cycle of research drives all campaign activities.

Through a nationwide network of trusted messengers and consistent, fact-based public health messaging, the Campaign helps the public make informed decisions about their health and COVID-19, including steps to protect themselves and their communities. The effort is driven by communication science and provides tailored information for at-risk groups. The Campaign supports efforts of the Centers for Disease Control and Prevention (CDC) and others across HHS to use education to improve public health.[10]

Under "Paid and Earned Media" we learn:

**Paid and Earned Media**
The campaign uses both paid advertising and media interviews, presentations, radio/TV tours, and other public events to educate people about the importance of vaccination and basic prevention measures to prevent COVID-19 and protect public health. The Campaign is designed to reach 90% of the American adult population at least once per quarter. On average, they would be reached at least 10 times.

The Campaign's paid media activities focus on three strategies:
- Slow the Spread—Provides general audiences with action steps they can take while waiting for the vaccine to protect themselves, family members, and their community, as well as tailored messaging to those who are disproportionately affected.
- Building Vaccine Confidence—Uses public education to build confidence about the COVID-19 vaccines so people are ready to get vaccinated when it is their turn. The timing of these vaccine confidence efforts aligns with increasing availability of vaccines.
- Preparing the Nation—Provides general audiences with information about vaccine development, safety, and effectiveness, including answers to common questions.

Earned media outreach features trusted messengers and influencers from multicultural and other communities hit hard by COVID-19 as well as experts from HHS agencies, the larger public health community, academia, and industry. They provide factual, timely information and steps people can take to protect themselves, their families, and their communities. This includes a new educational podcast, titled "COVID-19 Immunity in Our Community."[11]

## "Better Living Through Chemistry"—The Tragic, Forgotten Story of DES

Diethylstilbestrol (DES) is an artificial hormone synthesized by British chemist Edward Charles Dodds and colleagues—first published in the prominent medical journal *Nature* in 1938. This synthetic "nonsteroidal" estrogen was then synthesized by several different pharmaceutical companies around the world. Due to public funding of the research, DES was not patented. Dodds reported that DES could prevent pregnancy or induce abortion in rats and rabbits and consequently could be used as a contraceptive or emergency contraceptive pill. Dodds however, considered the possible adverse effects in women, and he censored the use of DES, suggesting that it might even induce cancer.

Concerningly—for profit no doubt—it was first used in women for the treatment of estrogen deficiencies for such things as menopausal symptoms, vaginitis associated with low estrogen levels, and the suppression of breast milk production after birth. DES was prescribed to pregnant women from 1938 to 1971, even though Dodd warned that it could induce abortions and prevent pregnancies in animals. For what reasons was it prescribed in pregnant women? Absurdly, to prevent miscarriages or premature births in the US, even though William Dieckman presented studies in 1953 showing that DES did not reduce them.[12]

The disastrous effects caused by DES are far reaching and very extensive including cancers of the uterus, cervix, vagina, breast, severe neurological abnormalities, and problems associated with socio-sexual behavior, as well as immune, pancreatic, and cardiovascular disorders. So called "epigenetic" alterations occur which, largely oversimplified, refer to inherited traits or change of cell function without changes to the actual DNA. DES was prescribed for much longer in other European and American countries, and lower-income countries constituted a major concern since DES was inexpensive, a factor that contributed to the use of this drug even after its withdrawal by the FDA.

In 1971, in the classic work of Arthur L. Herbst and colleagues, published in the *New England Journal of Medicine*,[13] DES was shown to cause a rare "clear-cell carcinoma," a rare vaginal cancer in those offspring that had been exposed to this medication while in their mother's womb. Worse, instead of preventing pregnancy loss, it caused a variety of malformations and hormonal disruptions in the preborns whose mothers took it during pregnancy. It was shown to cause a variety of significant adverse medical complications during the lifetimes of those exposed in their mother's womb, including congenital malformations of the cervix and uterus in females, vaginal adenosis, uterine fibroids, breast cancer, infertility, and depression. While DES sons were less known, their complications too were devastating. DES sons presented underdevelopment of the penis, un-descended testicles, infertility, lower sperm counts, and abnormal production of natural testosterone, intersexual defects, under-masculinization, and hypospadias—an abnormal development of the penis with the urethral opening located on the penile shaft, or penile base, instead of on the tip of the glans penis.

## DES Daughters: Multi-Generational Loss

Women exposed to DES while in their mother's womb who subsequently underwent pregnancy, experienced catastrophic complications including, miscarriage, ectopic pregnancy, preeclampsia, stillbirth, premature menopause, cervical insufficiency (weak, short, abnormal cervixes), premature rupture of membranes, premature labor, and neonatal death as the result of severe prematurity. Alarmingly, and especially relevant to the COVID-19 "vaccines," is that "third-generation DES daughters" experienced an increase in abnormal menstruation, absent menstruation, and preterm delivery. A study examining twenty-eight DES third-generation daughters also had well-documented genital tract changes associated with DES.

Experimental animal models—such as mice—are good for inter- and trans-generational studies (reproductive toxicity studies) because at least three generations can be observed and evaluated over the course of one year. Moreover, if DES multigenerational neoplastic effects observed in rodents also occur in humans, it will take about five decades to identify the effects in future generations. If this is the case, the story of DES is not over yet.

This will help the reader understand why I demanded long-term safety studies before the novel, experimental Covid gene therapies were imposed on the world. It helps us understand how irresponsible and insane it is for ACOG, ABOG, and SMFM to push these experimental gene therapies—these "vaccines"—in pregnant women.

## mRNA in Our Food Supply

DES was also used to induce growth in food-producing animals. This may sound eerily familiar to those who are aware of the recent injection of mRNA products into our animal food supplies, including pigs and cows. However, because of risks to human health, DES was banned in several countries since 1979. Unfortunately, the study by Loizzo documented high levels of DES in homogenized baby foods in Italy.[14] The results suggested that DES was subcutaneously or intramuscularly implanted into animals and remained in tissues for longer periods. This type of exposure of infants to DES could explain the increased detection of breast growth and precocious (early onset) pseudo-puberty in children. In addition, traces of DES have been detected in livestock, meat, milk, fish, shrimp, river water, and sewage treatment plants.

## Gross Negligence

Apparently, sixty thousand Ob-Gyn physicians lost their minds, with nearly all of them breaking the "Golden Rule of Pregnancy" by pushing the experimental gene therapies—these so-called "vaccines"—in their pregnant women. Did they have a gross lapse of common sense, memory, or lack of knowledge regarding the history of their specialty? Certainly, the threats from their professional organizations ACOG, ABOG, and SMFM contributed. And no doubt this "new age academia," better described as fascism, has put the focus on more politically expedient topics. Medical schools, residency training programs, the American College of Obstetricians and Gynecologists (ACOG), the American Board of Obstetrics & Gynecology (ABOG), and the Society for Maternal-Fetal Medicine (SMFM) were too focused on brainwashing the new generation of Ob-Gyns regarding the "new science of woke-ism." It was more important to teach them gender fluidity, that men can get pregnant and breastfeed, and of course, the all-important associated language of this mother-baby death cult that MUST be used, even in medical journals. The new woke fascism of the DEI doctrine (diversity, equity, and inclusion) drives contemporary obstetrical medicine. This is not medicine—it's political ideology!

How did "medicine" become transformed into a mere engine of "woke" revolutionary ideology?

Reliable sources inform me that the famed headquarters for ACOG have removed all the white, male, past presidents of ACOG, and everybody in the field knows that if you can't join the revolution, with letter-perfect virtue signaling, you'd better retire fast. You do not stand a chance.

# Chapter Seven

# Eight Candles: The Holocaust Hits Home

*Because an emotional massacre took place, in these little towns, across America. And now, we are expected to act as if this has never happened at all. Physically though—emotionally—there is blood flowing in the streets.*
—Naomi Wolf, *Facing the Beast: Courage, Faith, and Resistance in a New Dark Age*

The way they designed this evil, is like a mass die-off event that is both very obvious yet hides in plain sight. Much like the *Emperor's New Clothes*, in Hans Christian Anderson's famous fable, the clothes did not exist. But the tricksters, the ones who pulled the puppet strings, had convinced the whole village that only people who were stupid, low, and *unfit for employment*, would fail to see the emperor's clothes. Hence, everyone "saw" what was not there—except for one small child, who spoke the truth. We don't know his fate, as the original ending was removed. Many Anderson scholars speculate that the boy was killed for his innocent and truthful utterance.

People fear for their very lives when they go against powerful consensus. No typhoon of consensus and peer pressure had ever hit the United States before, carrying the velocity and sorcery of the Covid vaccine *psyop* of 2021. It shattered families and made it a social crime for family members even to warn one another, or to mourn their dead.

It's the collective blindness and deafness of a people, who refuse to notice, much less accept the crimes of their rulers, or the downfall of their civilization.

Everyone working in the medical system had been conditioned, without words, to simply never mention it. To see it, work around it, but never mention it. If they did mention it, if they did see it, like Lot's wife, they turn to salt. They're immediately deemed a threat in the workplace, bullied, frozen out, fired, and never hired again. Most in the system chose to just not say anything. Handle the dead babies, handle the dead bodies, fill out the paperwork, and go home.

For me, "seeing" it means constant torment. It finds me wherever I go, as soon as I wake up each morning; I'm back in the nightmare. I have not felt the same since this began, not felt at peace. I walk around with a searing rage I must always repress.

And every day, through my phone texts, and my emails, more horror stories come to light.

What follows is a story of my dear friend Nan and her husband, John, (names changed); a story about their son, daughter-in-law, and what happened to them in the years between 2021 and 2024.

Nan and John—nearing retirement age—live in a small town in Michigan and agreed to talk to Celia and me on a Zoom call in Spring 2024. Sitting on their sofa, with an abstract painting behind them, they were stiff with sadness.

The family's tragedy seemed frozen in a silent mountain that has absorbed all words and sound. It is not to be spoken of. But they are nevertheless choosing to speak to us, as they are old friends of mine, and Nan had alerted me to the tragedy earlier on in a Christmas card I received from her in December 2023.

Before Covid, their son and daughter-in-law had been blessed with a healthy child—her first grandson—a boy, now four years old. Everything was perfectly normal with their first child, with his conception, the pregnancy, and the birth.

It was 2021 when her daughter-in-law, Sarah, and son, Luke (names changed), tried for a second child. Sarah had been reluctant to take the Covid shots, but her Ob-Gyn persuaded her, despite the fact she was four months pregnant. She gave in.

A week later, Sarah miscarried. "Well, this can happen," she and her husband thought, "it wasn't necessarily because of the shots." By that time, she had received two Pfizer Covid vaccines.

The couple waited a few months before trying again. This time the miscarriage happened sooner, in the first trimester. Again, they brushed it off, as probably normal. But they felt a rising anxiety.

# Eight Candles

Prior to these failed pregnancies, Sarah had no problem getting pregnant. She and Luke were prepared to do whatever it took to have a second child, so they tried again. This time it was twins. They were cautiously elated and vowed to be extremely careful in every way.

Sarah carried the twins until late in the second trimester, when she suddenly lost her pregnancy, yet again. A boy and a girl. They had now lost four babies.

They tried again. The next baby was lost late in the first trimester, and the one after that, in the third trimester.

## Silence Imposed

The trauma of it all caused the young mother, now expecting again, to shut down all discussion of her pregnancies. Nan and her husband were sent a list of things they may not bring up. Nan felt she was at risk of losing this whole branch of her family, so naturally, she was compliant and obedient. She kept both her feelings, her grief, and her knowledge to herself. The family carried on and tried to make "normal" out of something extremely harrowing and out of place. The final end station is silence: The whole family just stops speaking of it altogether, from sheer trauma. This is known in trauma circles as the "freeze" response.

Meanwhile, a different kind of silence struck hospital workers. Doctors and nurses somehow created and encoded a pact of Omertà—a code of silence and secrecy sworn to by oath—and made sure that if any of them were reacting to the many miscarriages and stillbirths, they were not doing so audibly or showing any expression on their faces or through their body language. In this way, the mRNA holocaust went into a half dream-space where people could only mourn in top secret, in quiet corners of their minds. But never out loud, or to each other.

Nan's daughter-in-law, like so many women in her position, created a wall around herself, just to survive.

After the loss of six babies, Sarah wrote a letter to the family, stating that she wanted nobody to ask her *if* she is pregnant, nor ask any questions *about* her pregnancies, nor speak of it *at all*. She would alert the family when she had given birth to a living baby.

"It was like a manifesto," said Nan, in our Zoom conversation. "[The letter] had a list of rules. This dramatically changed our family dynamic. We rarely saw them anymore. It felt like they didn't want us around. Like they were afraid we would say something about why we thought this was happening."

Nan, whom I've known for over forty years, knew of the dangers of the shots because of my warnings, and had conveyed her worries to her son and daughter-in-law. Unlike most of my friends, they had watched my interviews and heeded my warnings, instead of just assuming I had "lost it."

"I told them not to do it, not to take the shots," she said.

Despite Nan's warnings, her son and daughter-in-law went the "normal" route and followed the CDC's edicts for disaster: There was no inopportune time to take the Covid shots, a smiling Rochelle Walensky assured the nation from TV screens where she was suddenly ubiquitous. So, Sarah—who had a completely normal pregnancy and birth *before* the vaccines—got one Covid shot before she tried to conceive again, punctuated by at least two boosters.

Then the carnage began.

## A Family, Fractured

Though she had nothing but love and empathy for her son and daughter-in-law, Nan felt herself more and more shunned, as though she carried a Cassandra curse. She grieved each lost grandchild quietly, off by herself, or in shared sorrow with her husband.

But she also grieved the loss of her formerly happy, cheerful daughter-in-law, who grew more and more withdrawn, finally all but ceasing to speak to her altogether, or to reach out.

"I didn't even get a text from them on Mother's Day. And when I sent gifts, I got no response. It was as though we had all already died."

Her heartbreak continued. Sarah lost two more babies to miscarriage—one in the third trimester, and one, a boy, at birth. The boy had already been named Gabriel. He had been fine, just the day before, with a normal heartbeat, no abnormalities at her hospital checkup. The staff had been monitoring this pregnancy *very* carefully, because of Sarah's history.

Sarah was at her due date, and suddenly felt nauseous. She and Luke got to the hospital just in time to save her life—her placenta had completely separated from the uterus internally (placental abruption). But Gabriel was dead; he had probably died moments after the rupture, which cut off his oxygen. Before it was all over, Sarah had lost as much blood as a human being can lose, without dying.

Who can imagine such a thing, for a young woman, who had already lost *seven* babies, before this? And on TV, every night, they were telling America's pregnant women there was, to quote CDC's Rochelle Walensky, ". . . no bad time to get the Covid vaccine."

# Eight Candles

On Sept. 29, 2021, to cite another one of countless such repeated lies, Walensky was quoted on CNN saying:

> We now have data that demonstrates that vaccines—in whatever time in pregnancy or lactating that they're given—are actually safe and effective and have no adverse events to mom or to baby.[1]

"Her dreams are shattered," says Nan. "Sarah is no longer the happy, fun person my son married. She is almost catatonic at this point. We are all shell-shocked, unable to talk about it.

"They held a funeral for Gabriel, the last baby they lost. The other babies were also memorialized, each with a lit candle, which Sarah and Luke blew out, one by one. Eight dead babies. It was the most heartbreaking thing I've ever experienced," said Nan.

She went on, "My beautiful daughter-in-law, and my son, they lost eight babies, all told, in all three trimesters, including the third, which was the last two. The child had no chance, and we are lucky we didn't also lose Sarah. She came very close to dying. They'd never seen anybody lose that much blood and survive.

"You have to understand something: There is no way to talk about this, in afflicted families. Instead, we just turn away from each other, rather than confront it, or let it be mentioned. It's too much for any psyche to bear.

"Our world is upside down, and evil has been unleashed on God's people," Nan said quietly, looking down.

At that point in our Zoom call, her husband, John, placed his hand on hers. They looked shipwrecked, on their sofa, floating in a cold cosmos, or dark ocean—a foreign land, where nothing was recognizable. How could they ever piece their family back together again?

I finally spoke. "It's not a normal pattern." A young woman who gave birth to a healthy child does *not* have eight pregnancy losses including late miscarriages, fetal deaths, and a newborn death after that, in all three trimesters. None of this is "normal" or "imaginable."

"She's probably too terrified to try again," Nan sadly shared.

"And, at this point," says John, "we're just hoping she can stay alive—literally. She had overcome episodes of depression before this, and we are very concerned about her mental health. But we can't get close to them. We have no idea how to help them."

My old friend knew she was safe unburdening her heart with me. In the Christmas letter she wrote, "The evil forces in this world want to control population, . . . Promoting abortion wasn't enough . . . the vaccine has destroyed many more."

She's right. It's hard to outstrip abortion but these shots will, when all is said and done, kill more babies—wanted and welcome babies, making it all the more tragic and sad.

It's especially the later miscarriages and fetal deaths that are hard to stomach. Thinking about the expectant couples, the mother being so attuned to that life within her, and then out of the blue, a serial killer strikes, and it's clearly not "Mother Nature." How many old fairy tales revolve around an evil potion, or the dangers of alchemy, and sorcery? This wicked stuff is actually, to state it simply, an overt attempt to overwrite God's plan and blueprint for life. They're not trying to stop or avert death; they're trying to replace God and become the new god, with all of humanity enslaved and controlled by them, by their destructive and unnecessary "vaccines."

It's worth noting here, that the word "sorcery" is used in the biblical Book of Revelation four times, in Chapters 9:21, 18:23, 21:8, and 22:15. In Revelation, Chapter 18 verse 23 John, speaking of the end of time, wrote the following:

> . . . all nations were deceived by your sorcery.

In the context of the Bible, "pharmakeia" is often translated as "sorcery," "witchcraft," or "sorcerer." Think about that, just for a moment. So often, the truth of what we need to know, is just one curious click away.

The following passage from Acts, when I reread it closely, sends a chill down my spine. These babies are dying because Godless atheists decided to issue a blasphemy so grave it escaped even devout Christians in churches across the country in 2020: The "vaccines" deigned to "fix" what was made by God but was not even broken. They deigned to impose their own synthetic "mRNA" blueprint of life over that of our Creator and they were insane enough to think they would be seen, therefore, as gods. And for a short period of time, they were. They had their worship. But then the deaths began.

I get a tear in my eye when I read these simple words:

> *Nor is he served by human hands, as though he needed anything, since he himself gives to all mankind life and breath and everything.*
> —Acts 17:25

# Eight Candles

Life and breath and everything.

The Covid vaccine crime outstrips any other crime on earth. Any victim of any other crime can at least admit what happened to them, and central to that admission is the healing process, as Elizabeth Kubler Ross outlined in her writings about the stages of grief. Victims of this crime will have little chance to know, to grieve, or to recover from what befell them—nobody has admitted the truth; they can't admit it to themselves, and entire families are being engulfed by these traumatized silences.

Michelle Spencer, nurse turned whistleblower, had shared with Celia and me that most of the women who were her patients—who lost their babies due to Covid shots—tended to turn the blame on themselves. "They thought it was something *they* did. They thought it was because they had high blood pressure, or something like that. They definitely tended to blame themselves.," Spencer said.

But the truth was evident in the vaccine manufacturers internal documents the whole time, clear as day. In her 2023 book *Facing the Beast: Courage, Faith, and Resistance in a New Dark Age,* Naomi Wolf writes:

> The volunteers found that while pregnant women were excluded from the internal studies, and thus from the EUA on which basis all pregnant women were assured the vaccine was "safe and effective," nonetheless about 270 women were reported as pregnant in Pfizer's post-marketing report. More than 230 of them were lost somehow to history. But of the 32 pregnant women whose outcomes were followed—*28 lost their babies.*
>
> That's more than 80 percent.[2]

In a later passage, Wolf emphasizes this: "They targeted the human fetus' very environment, one of the most sacred spaces on this earth, if not the most sacred."

## "My Body, My Choice"—Except in the Case of COVID-19 Vaccines

Dr. Naomi Wolf stands almost completely alone in a field once populated by the robust tribe she used to identify with: feminists, by definition, "on the left." A half century after the imposition of the psyop known as "feminism," Dr. Wolf could not rally *a single one* of her old feminist allies to utter one peep against this carnage against women and their babies. Instead, she became the object of public scapegoating and vituperation, culminating with a grotesque attack from "the other Naomi,"—namely, Naomi Klein—who stalked Wolf online, and penned a creepy pop thriller pretending to trace how Naomi Wolf had lost her mind.

Klein's 2023 book, *Doppelgänger: A Trip into the Mirror World*, was such an advanced piece of gaslighting, it denied all the carnage and loss to women and babies who fell victim to the Covid shots, and shone a spotlight of shame on Naomi Wolf alone, for decrying it. What happened to the rallying cry of "My Body, My Choice"? The hypocrisy was so utterly blatant, for those who could "see."

But such was the power of conformity and denial in the Covid era— that's how degraded and insignificant women and their rights had become, from those who purported to defend them.

In the words of Alexander Solzhenitsyn:

> *The simple step of a courageous individual is not to take part in the lie. One word of truth outweighs the world.*

# PART TWO

## Chapter Eight

# The Obsolete Man

*This is not a new world. It is simply an extension of what began in the old one. It has patterned itself after every dictator who has ever planted the ripping imprint of a boot on the pages of history since the beginning of time. It has refinements, technological advances, and a more sophisticated approach to the destruction of human freedom. But like every one of the super-states that preceded it—it has one iron rule: logic is an enemy and truth is a menace.*
—Rod Serling, *Twilight Zone*, "The Obsolete Man"

"The Obsolete Man"—which aired on CBS in 1961 as episode 65 in the famous series—is about a librarian in a future dystopia who was put on trial by a militantly atheist totalitarian state for being "obsolete." The character, modeled on George Orwell's Winston Smith, was called Romney Wordsworth.

In that episode, Romney is depicted as "obsolete" in every regard. But mostly, he has been deemed obsolete for loving books and for believing in God. Books were banned and the state mandated that all citizens accept that God did not exist; the state had supplanted God.

Romney Wordsworth is tried and sentenced to die for his alleged disobedience, but he is permitted to choose the manner of his death, which would be broadcast on live state TV. He asks the prosecutor (chancellor) to be his executioner—a wish he is granted—though the manner he chooses to die is not immediately revealed.

Alone with his official killer, Romney reveals he intends to die by a bomb going off in the room, and that incidentally, the door is locked, so the chancellor will also die with him. He then proceeds to sit down and read out loud from his Bible, Psalm 23:

> *The Lord is my shepherd; I shall not want. He makes me lie down in green pastures. He leads me beside still waters. He restores my soul. He leads me in paths of righteousness for his name's sake. Even though I walk through the valley of the shadow of death, I will fear no evil, for you are with me; your rod and your staff, they comfort me. You prepare a table before me in the presence of my enemies; you anoint my head with oil; my cup overflows. Surely goodness and mercy shall follow me all the days of my life, and I shall dwell in the house of the Lord forever.*

The chancellor, in abject panic, begins to plead for his life. Romney releases the chancellor from his fate and spares his life. The bomb detonates; Romney dies.

Following this humiliating exposure of cowardice, the state replaces the chancellor with his "subaltern," and he is soon beaten to death.

Serling's stunning conclusion, in a rare closing monologue, was this:

> The late chancellor was only partly correct. *He* became obsolete. But so was the state, the entity he worshipped. Any state, any entity, any ideology, which fails to recognize the worth, the dignity, the rights of man, that state is obsolete.

We are all (this includes all who are reading this) in an existential battle with a malicious totalitarian state that descended upon us, seemingly out of nowhere, in the late winter of 2020. I call it the Covid State. With the absurd pretext of a coronavirus from Wuhan, China, fast becoming a global "pandemic," all aspects of civil, economic, social, and certainly medical life as we knew it, were pulverized. It was as though a silent, massive detonation had occurred. When it was over, we were in terra incognito; we no longer recognized our world.

Each of us found ourselves, in the waters of post-blast, clinging to a shard of something we hoped would float. The blast in my world as a physician, left me without walls or a roof on what had been the "house" of obstetrical medicine itself. When I looked at my world after the blast, the Golden Rule had been obliterated. Not only that, it was as though the Golden Rule

# The Obsolete Man

had never existed. Everyone and everything that did not fall in lockstep with the new Covid agenda and revolution, became obsolete.

If you are reading this, you too, are "obsolete," as am I—since we object to the Covid (Beast) State and can see its true face. By buying and reading a book such as this, you're already an enemy of the lie, and therefore, of the state.

The "Beast" itself is doing all of this, I believe, because it is programmed, ultimately, to destroy itself. This is why, as dark and unfathomable as things are, I want to prepare you that while reading this often-depressing text, I assure you this evil entity will fall. Why do I say that? Because, as Serling correctly wrote: "Any state, any entity, any ideology, which fails to recognize the worth, the dignity, the rights of man—that state is obsolete."

The State will appear to be omnipotent, but it never is. It's inflicting evil on borrowed time, like any schoolyard bully. There is no dictatorship in history that has survived for long, but while they are perpetrating their crimes, they appear invincible. This particular dictatorship is unusual in that it demands the compliance of the entire medical profession to carry out the state's commands for covert executions. This also means that doctors, not known for heroism, as such, were given a choice to either betray our oath—the Hippocratic Oath—or, be professionally terminated.

The Beast attacks all vestiges of the truth about itself and tries to destroy anybody who exposes it, no matter what credentials they may have. While achieving board certifications in both Ob-Gyn and in sub-specialty maternal-fetal medicine (high risk obstetrics), for over forty-five years, I have focused on the protection of my patients: pregnant women and their unborn children.

There came a day when the medical system I always believed in and served loyally and fruitfully, all my career, terminated me, and told me in tones of high praise, that I had become obsolete. They did not use that word, but that's what it was. I realized I had been living in a fictional reality all my life—a reality where these things did not happen in America, only in other countries. Countries that are not fortunate enough to have our Constitution.

But it did happen here. It happened here on a scale that cannot ever be described, no matter how many books are written about it. Our government institutions became openly homicidal. Even now, as I reflect back, I think to myself: "It's not possible. It did not happen. I must be in some kind of very long, very strange dream."

## Chapter Nine

# Getting Fired: A No-Cause Termination with a Cause

*"Then we must go as we are," said Ralph, "and they won't be any better."*
*Eric made a detaining gesture, "But they'll be painted! You know how it is."*
*The others nodded. They understood only too well the liberation into savagery that the concealing paint brought.*
—William Golding, *Lord of the Flies*

I had been speaking out, you might say, "broadcasting," to anyone who might listen about the deadly Covid shots since summer of 2020. This included scores of interviews, podcasts, and testimonies, most notably, an interview with Tucker Carlson—while he was still at FOX—and the US Senate Testimony Roundtable arranged by Senator Ron Johnson, on December 7, 2021.

I had also tweeted out 7,849 communications on X, formerly Twitter, to my over 55,000 followers. I didn't soften or sugarcoat anything—I just made sure I was being medically and scientifically accurate. Plainspoken by nature, I'm also an open book. I don't know how to play at politics, and I'm not interested in learning that game. After all, what is a job, or a reputation, compared to a grotesque crime against humanity of this scope?

There are so many physicians in this post-American landscape who, just like me, didn't "get the memo." We saw medical murder and we called it medical murder. There was never any time to think, ponder, or make decisions. We spoke out. We acted. So many of us have been stripped of our medical licenses, fired, targeted, smeared, harassed, bullied, mocked, and in some cases I suspect, even taken out. Those were the threatening waters we swam in. I almost wondered why I had still not been fired by the summer of 2023, but I had sensed a kind of darkening shadow approaching—a sword of Damocles.

On June 29, 2023, it finally came down, but not swiftly, as I would have preferred.

## An Unusual Conversation

I was scheduled to speak to Kevin Elledge, the CEO of SSM Sisters of St. Mary's Health System, where'd I'd been employed for five years. He had sent an email to say he wanted to talk to me. At the agreed upon time, he called.

Once we were on the phone, the best word I can use to describe the conversation is "bizarre." Like everything in Covid-land, nothing made sense; B did not follow A. Nothing was recognizable from the world I knew before Covid.

The conversation was not only strange, it was also unusually long, and that made it take on a slightly surreal feeling. For the first thirty minutes or so, we exchanged seemingly friendly "small talk." We spoke about our lives, our families, vacations we'd like to take, and the like. After that, Elledge began to praise me, more and more fulsomely, and conveyed a variety of compliments he had received from my colleagues, superiors, and staff.

Finally, a half hour in, the hammer dropped:

"I have some bad news to share with you Dr. Thorp," he said. I was quiet, listening carefully.

He began to speak of financial difficulties at SSM Health, telling me they'd lost $23 million in the month of May alone, partly because of rising nursing agency costs. That—dear reader—was a head scratcher.

"Covid," as we know, was a typhoon of profit for the entire US healthcare system. The HHS allowed a total Covid budget of over $5 trillion, and of that, billions went to the hospitals, which were heavily incentivized to see, diagnose, and "treat" Covid, even if somebody had arrived dead from a motorcycle accident.

And then, rather abruptly, he said it: "Dr. Thorp, the decision was made to end our relationship with you." After a brief pause, he added, "I know that must come as a shock."

Elledge then said something I found memorable, and curious.

"I can't even understand why I am even saying this to you," he said, as though confessing he was partly operating from some kind of trance state. I felt that he was very uncomfortable—twisting and turning, trapped in the gears of this frightful hospital killing machine, which was spitting me out, just as it was tightening its grip on him.

At times it felt like he needed more consoling than I did.

"You are such an outstanding, dedicated, and loyal physician to our system," he carried on. "You are everything we would look for and a model physician."

He still wasn't finished telling me how wrong this all was *at heart*.

"Companies tend to make crazy decisions when they are in a crisis, and I'll be the first to admit that this does not appear to be in anybody's best interest."

I just let him talk.

Finally, he got down to the matter of my contract. "Our current agreement," he said (signed October 18, 2018), "contains a 'no cause' termination with a 120-day notice, that could be invoked by the employer or by the employee." He added that it was possible that in a couple months' time, they might bring me back on board if their financial situation improved. Next, he pointed my attention to an email he would soon be sending, containing an important letter he wanted me to take my time and read. Then, he said he was sorry.

When I at last spoke, I said: "Is there something else that has gone into this decision other than what you are telling me, other than your financial issues?" I repeated that question three times during the remainder of our conversation. "NO, not at all," he insisted. "This is a 'no-cause' termination, it's not a 'for cause' termination or anything like that." He kept calling it a "business decision." He then shared the names of three individuals who were responsible for making this "business decision": Dr. Mary McClennon, Chairwoman of the Department of Ob/Gyn; Dr. Gil Gross, Director of the Division of Maternal-Fetal Medicine; and Donna Spears, Administrative Director of the Women's Care Line of Service. He repeated that they all thought very highly of me and called it a "difficult, difficult decision to make."

To my surprise, he then told me they wanted to give me $90,000 if I signed the "separation agreement"—the letter contained in the email he had sent while we were on the phone. While I hadn't yet read it, I knew that a separation agreement would include silencing me, so I said essentially "Thanks, but no thanks."

My instinct was correct. Hours later, when my in-house counsel, Maggie, reviewed the separation agreement, she pointed out that it included the most threatening non-disparagement clause she had ever seen. Not only was I not

# Getting Fired

allowed to "disparage" SSM Health, but I was also responsible for anyone else independently criticizing SSM Health for what they did to me regarding my termination or anything else.

I repeatedly told Elledge that I was unwilling to take their generous money award but appreciated the offer. "I feel very uncomfortable taking any of your monies, Mr. Elledge, because you just told me that SSM Health is having financial difficulties. Please keep your money."

What I did not share, was that I would *never* accept that money because I would consider it blood money, deployed to buy my silence.

Elledge then became much more aggressive, almost demanding I take the money: "You said you don't want or need the money," he said, "but please take it for one of your five grandchildren. They could certainly use it."

Respectfully, I dug in my heels and told him that under no circumstances would I take it. I told him that I had an incredible experience with SSM Health, the administration, physician staff, nurses, sonographers, and everyone that I worked with. "I will not take your money," I said. "I feel uncomfortable accepting it."

While I knew I deserved that money, having been the most productive physician by far in the Division of Maternal-Fetal Medicine—and likely in the entire Department of Ob-Gyn—I also knew it was tainted. I would be damned if I took it. It would have been a betrayal of God, myself, my Officer's Oath to uphold the Constitutional First Amendment, my Hippocratic Oath, and the sanctity of my physician-patient relationship.

While Elledge became increasingly frustrated that I refused the $90,000 separation payment, which I considered a bribe to buy my silence, he maintained his respect for me. The conversation ended at 1:50 p.m., on June 29, 2023. Exactly seven minutes after we had hung up, the SSM Health System sent out a mass email to everyone I worked with informing them that, effective immediately, I was terminated and would not be seeing any more patients.

I was shocked of course, as was Maggie, because Elledge had just made it very clear that I was being terminated for "no cause" and could work out my clinical responsibilities and continue the care for my colleagues and patients for 120 days. This was very upsetting, because everyone reading that email would realize that nobody is ever immediately terminated without cause, and it would be a clear assumption that I was a horrible physician and had done something terribly wrong. In addition, SSM Health immediately terminated all my telemedicine accounts and platforms along with them.

That very afternoon, I received multiple communications from my stunned colleagues.

All the SSM employees that contacted me conveyed that Elledge and SSM Health were lying. There were absolutely no other employees at any level in the Women's Line of care that were laid off, fired, or furloughed. This would have been expected if the real reason for my termination was caused by "financial distress," as he so claimed.

My colleagues went on to comment that I was the number one income producer for the Division of Maternal-Fetal Medicine and likely for the Department of Ob-Gyn. This is accurate, as I kept track of my clinical activity and had seen over 27,500 high risk obstetrical scans in about 4.5 years with SSM Health. This number exceeded all the other physicians, as was previously brought to my attention by my Division Director over a year ago. Unfortunately, and undoubtedly, my reputation was severely damaged by the mass email that was sent out that day. And not only did they slander my reputation, but SSM Health refused, spitefully, to pay me for the 120 days—in breach of contract, no less.

I knew deep down, that somehow, God would take care of my needs and the abrupt loss of income. Indeed, He did. Within seven days I had exactly as many offers for new employment.

## The Lies Persist

I posted a Tweet[1] on August 21, 2023, about seven weeks after my termination from SSM Health on June 29, 2023, which stated the following:

> SSM is this the result of the $306 MILLION you received from HHS CDC? Apparently, you sold your souls for money. @SSMHealth
>
> @SSMHealthSTL #LauraKaiser #KevinElledge You owe your employees and patients AND the American Public clarity on the contract you signed w HHS CDC. It was our tax money. Appears as though you entered into a quid pro quo agreement with the HHS CDC signing the "Covenant With Death" exactly like The American College of Obstetricians & Gynecologists (ACOG) did. You appear to be threatening physicians & employers with termination for not following the lethal narratives of HHS CDC. Horrifying. Anyone else in your system as extensively published on COVID-19 as I am? Why was this not addressed in an academic manner? This appears more consistent with fascism power and money—certainly NOT science.
>
> SSM HEALTH Employees. Be Bold & Courageous. Stand up for truth. Silence is complicit. SSM patients- ask your doc, nurse, tech's & administrators if they are willing lose their job to save your life or your babies life. Unfortunately it appears the vast majority will not.

# Getting Fired

This Tweet[2] included a picture of an SSM advertisement showing a young Dr. Shephali Wulff an "SSM Health Infectious Disease expert" stating "The data is indisputable. If you get the vaccine, you are less likely to develop severe disease or die as a result of the virus. At this point, the vaccines have been well-studied and they are safe." This was known at the time to be categorically false, proven by multiple data sources including the CDC/FDA VAERS data and by Pfizer's own data that it was anything but safe; it was the deadliest, most injurious drug ever rolled out to the public. (see pages 19, 20, and 22 that were also pictured in this Tweet). Multiple sources actually suggest the exact repeated vaccination increases the likelihood of re-infection with COVID-19 (negative efficacy).[3, 4]

As of 11/20/2024, despite this post being shadow-banned, it has 146,100 views, 1,400 likes, and 946 retweets. I intend to keep this tweet pinned to the top of my account until justice is served to SSM Health for the likely deaths and injuries they have caused, to their employees, and the entire population of patients in their service area. SSM has violated the first amendment rights of all their employees by entering into a contractual agreement with the HHS and CDC, gagging free speech with punitive measures.

On a more positive note, because of this firing, I found myself in good company. An astounding number of American physicians have been either harassed, terminated, de-licensed, sued, censored, and de-credentialed by their specialty boards. I refer to these heroes as "Semmelweis-ian Heroes of this day."

Why? All through medical school, I would imagine being able to talk to the man himself, though he was long dead. Dr. Ignaz Semmelweis, of the University of Vienna Medical School, greatly influenced my interest in obstetrics and in independent clinical research. It was the mid-1800s, when Semmelweis underwent harassment—and even worse—for speaking the truth. Back then, physicians were inadvertently killing their patients—birthing mothers (up to 50 percent in some months). They were unknowingly infecting them by going from the autopsy room straight to the delivery room without first washing their hands, as that was not yet an established practice. As you'll read in a later chapter, Semmelweis died at the young age of forty-seven, after those who did not "approve" of his findings—his truth-telling—forced him into a psychiatric ward, where he succumbed to sepsis.

## Weaponizing "Mental Illness"

One of the mechanisms used to harass and coerce physicians in this era of totalitarian medicine is the classic Soviet tactic of declaring us "disruptive" and suggesting we are mentally ill and need psychiatric counseling. That's precisely how they ensnared and destroyed Dr. Semmelweis, back in the 1800s.

The powers-that-be threaten these measures as protection against being fired—telling us we will be terminated if we refuse. Certainly, there are disruptive physicians whose behaviors are inappropriate. I am not debating that. But this strategy has been weaponized to control physicians whose only "mistake" was to speak the truth and defend their patients against the violent state, with its deadly protocols.

I realize now, that despite my getting fired, speaking out was a courageous act, although my conscience would allow for no other choice. It was as though my whole career practicing medicine has been in preparation for this moment, when I would be called to demonstrate that I take the Hippocratic Oath not only seriously, but also for granted.

It's so simple really, and now I see more clearly, how sacred these four words actually are.

They sound to me, now, like a simple, but powerful prayer:

"First, do no harm."

## Chapter Ten

# The Threat: ABOG's Attempt to Silence Me

*Threats betray the speaker by proving that he has failed to influence events in any other way. Most often they represent desperation, not intention.*

—Gavin de Becker

In 1985, after a written American Board of Obstetrics & Gynecology exam, and an in-person four-hour oral exam by multiple testers, I was permanently granted lifetime board certification in Ob-Gyn from ABOG. In 1991, after yet another set of both written and oral examinations, I became board certified in the subspecialty of maternal-fetal medicine (high-risk obstetrics). This required a published peer-reviewed medical journal article that I needed to defend. At the time, this was a very rigorous process, but over the years has been diluted in its intensity.

Fast-forward to 2021 after the rollout of the COVID-19 "vaccines," when I observed a surge in pregnancy-related adverse outcomes including heavy bleeding episodes, miscarriages, infertility, stillbirths, fetal malformations, severe post-partum hemorrhages, and many other issues. Indeed, I was seeing exactly what the Pfizer 5.3.6 document data had revealed on page 12. I had already done extensive research in the governmental database, the so-called Vaccine Adverse Event Reporting System (VAERS), that verified all my clinical observations. At this time, I had been in practice at SSM Health, St. Louis, and was the most experienced and the highest volume maternal-fetal medicine practitioner in my division.

In the Fall of 2021, I received a shocking email from ABOG, copied below.[1]

**Statement Regarding Dissemination of COVID-19 Misinformation**
*September 27, 2021*

The American Board of Obstetrics and Gynecology (ABOG) fully supports the statement published by the Federation of State Medical Boards (FSMB)[2] that asserts that providing misinformation about the COVID-19 vaccine contradicts physicians' ethical and professional responsibilities, and therefore may subject a physician to disciplinary actions, including suspension or revocation of their medical license. Additionally, ABOG supports a recent American Board of Medical Specialties (ABMS) statement,[3] which expresses concern regarding the serious public health effects of the persistent spread of misinformation regarding the COVID-19 virus.

Patients rely on physicians to practice evidence-based medicine based on facts and scientific data. The FSMB and ABMS statements align with the ABOG standards and policies for certification and maintenance of certification that involve medical professionalism and professional standing. These standards include:

- acting in your patients' best interests
- behaving professionally with patients, families, and colleagues across health professions
- taking appropriate care of yourself
- representing your Board certification and MOC status in a professional manner

Providing intentional misinformation that may harm patients or public health does not meet these standards and may be grounds for adverse action on OB GYN certification status. In addition, ABOG is alerted if an OB GYN is investigated for practice or professionalism violations by a state medical board. ABOG acts based on its certification policies as described in the current MOC Bulletins and the Revocation of Diploma or Certificate Policy.[4]

A recent article[5] highlights the risks posed to pregnant people by COVID-19 and the reports of increasing numbers of unvaccinated pregnant individuals in ICUs in several states with severe infection. The CDC has reported[6] that the deaths in August are the highest number of deaths reported in any month since the start of the pandemic, citing that about 97 percent of pregnant people treated in the hospital for COVID-19 have been unvaccinated. The ACOG[7], SMFM7, and CDC[8] recommend the COVID-19 vaccine for all people over the age of 12, including pregnant people. ABOG supports these

> recommendations and has incorporated this information in our Maintenance of Certification (MOC) learning and self-assessment offerings to help diplomates provide evidence-based care to the people and families that we serve.

This communication was upsetting on so many levels that my emotion quickly turned to anger. I had never witnessed such a repressive and demeaning fascist dictate from a professional organization in my life, let alone from ABOG who I had, until now, respected.

Given what I was experiencing, the board's lofty description of "why they exist" began to feel Orwellian to me. On their website, it states:

> Our mission, is to define the standards, certify obstetricians and gynecologists, and facilitate continuous learning to advance knowledge, practice, and professionalism in women's health.[9]

By what measure were they defining standards while ignoring the fact that babies of vaccinated women were dying at alarming rates? Clearly, in the Covid era, advancing "knowledge, practice, and professionalism in women's health" had been shoved aside to make way for their new mission: eliminating "misinformation" from their members' offices around the country. So serious was this crime of "misinformation" that one's board certification and their medical licenses would be revoked if one were "caught." But here's the irony. Nowhere in their "Statement Regarding Dissemination of Covid-19 information" did they define or provide examples of this dreaded "misinformation."

It seems logical, that would be clearly defined at the top of their statement, but instead, the statement begins with a threat:

> The American Board of Obstetrics and Gynecology (ABOG) fully supports the statement published by the Federation of State Medical Boards (FSMB) that asserts that providing misinformation about the COVID-19 vaccine contradicts physicians' ethical and professional responsibilities, and therefore may subject a physician to disciplinary actions, including suspension or revocation of their medical license.[10]

The FSMB's statement begins with a similar threat:

> Physicians who generate and spread COVID-19 vaccine misinformation or disinformation are risking disciplinary action by state medical boards, including the suspension or revocation of their medical license.[11]

Again, no definition or examples of "misinformation." They're not saying what this misinformation is, only that it's bad, very bad, and if you get caught doing it, you will be ruined.

The ABOG's statement also references the American Board of Medical Specialties [ABMS] statement supporting their concern about misinformation. No definition or examples of misinformation there either, neither from ABMS or from the statements of the other majestic medical boards to which ABMS provides links: The Federation of State Medical Boards, the American Board of Emergency Medicine, the American Board of Pathology, the American Boards of Family Medicine, Internal Medicine and Pediatrics, and the American Medical Association.

All strongly deplore "misinformation"—and that's it.

The next ABOG statement links to the "Revocation of Diploma or Certificate," reminding all "mis-informers" of the dire consequences awaiting them. Then, it provides a link to a *MedPage Today* article by Kara Grant, "More States Seeing Uptick of Pregnant Covid Patients in ICUs," and subtitled "Nearly all are unvaccinated, sources say."[12] The stunning fact is that ABOG presents zero peer-reviewed data and ridiculously cites a lay journalist writing for a magazine on par with a grocery-store tabloid.

It appears clear that the government, CDC, FDA, and others considered those patients in the Grant article to be "unvaccinated" for two weeks after their first shot, at least as far as efficacy was determined; but many allege that this was a manipulated definition that was also used with regard to adverse events. This is an extremely crucial point because it is now common knowledge that 40 percent of the deaths after the vaccine occur within forty-eight hours, and likely up to 80 percent within fourteen days.

The lie was propagated that pregnant women were at much greater risk of dying from COVID-19, which was proven false by Dr. Beth Pineles,[13] a maternal-fetal medicine physician who documented in a very large peer-reviewed study that pregnant women had a 75 percent reduction in dying from COVID-19 compared to non-pregnant women.

The September 15 Kara Grant article goes on to cite her previously published study from the month prior in *MedPage Today* (August 14, 2021) entitled, "'Alarming' Number of Pregnant Women Admitted to Alabama ICUs."[14] Grant makes a strong push for all pregnant women to get vaccinated. She cites physicians, suggesting that the University of Alabama at Birmingham had thirty-nine pregnant women with COVID-19 admitted to their hospital over the month of August and that ten were on ventilators. The article stated that "most were not vaccinated or not fully vaccinated;

some had received their first doses within the last couple of weeks before hospitalization." Really? What were the real numbers and how many really failed this miraculous vaccine that was falsely advertised as being 95 percent effective by Francis Collins, then director of the NIH (National Institutes of Health), just months earlier?

The August 24 Grant article[15] states that "the CDC strongly recommended that pregnant women get vaccinated against COVID-19 in the wake of new safety data." What safety data? Pfizer's data available months earlier documented it was the deadliest and most injurious drug ever, with 42,086 casualties including 1,223 deaths and an 81 percent miscarriage rate in just the first ten weeks after it was rolled out to the public (December 14, 2020 to February 28, 2021). Neither Grant nor any of the physicians interviewed, acknowledged the obvious conflict of interest: The University of Alabama is essentially an extension of the NIH having received over $325 million in 2020,[16] and an additional $610 million in 2023.[17] It's all about money. It is a well-known fact that researchers' results, and physicians' opinions, are highly influenced by those who fund their research or pay their salaries. He who pays the piper calls the tunes.

So here, finally, is the first indirect allusion to what the dreaded "misinformation" might be, via an article that mixes facts and declarative statements so lacking in context, that they can easily be construed to make a case for women to NOT receive the vaccination. Unfortunately, it fell short. The subtext message to pregnant women in this article is: be afraid, be very afraid, to be unvaccinated.

## Early Treatment Protocols Maligned

The sin of omission in writer Kara Grant's reporting about these pregnant Covid patients is this unanswered question: How many of those unvaccinated women had access to safe, early-treatment protocols after they tested positive for Covid?

The answer is ZERO.

Protocols that included crucial early treatments with vitamin D, vitamin C, Zinc, iodine, and other safe and effective nutraceuticals in addition to repurposed prescription drugs that are widely known to be safe and effective—such as hydroxychloroquine and ivermectin—were villainized to the point of implying criminal conduct. It is well documented how the Covid Beast system falsely impugned every modality including early therapy that could have treated COVID-19 with a cure rate of over 95 percent.

In May 2020, *The Lancet*, one of the most "prestigious" medical journals in the world, published a massive global research paper that had been completely falsified.[18] The article alleged that hydroxychloroquine was dangerous and ineffective. Five months later it was retracted. Unfortunately, the retraction was completely buried after the damage had already been done. How many victims were told by their doctors to "go home and rest" when their symptoms first appeared? How many of those patients went home untreated until a cytokine storm brewed in their lungs, forcing them to go to the hospital where they died needlessly for lack of early care or were killed with remdesivir?

I have successfully treated large numbers of unvaccinated pregnant women for Covid who went on to deliver perfectly normal babies. The manipulative reporting in the Grant article, the mainstream media, the falsified *Lancet* article (later retracted) is, to my mind, grossly unconscionable because they're encouraging pregnant mothers to inject poison into their bodies only to inadvertently share that poison with their unborn babies.

In Grant's September 15 article (referenced above), in the second paragraph, she references Mississippi state medical officer Thomas Dobbs stating: ". . . on September 9th, eight pregnant women died from Covid-19 in recent weeks; their babies were all delivered prematurely and survived."[19] In my opinion, those mothers sacrificed their lives to save their babies from the deadly vaccine. In a large cohort of 1,062 pregnant and 9,815 non-pregnant women, all hospitalized with COVID-19 and viral pneumonia, Pineles found that pregnancy had a dramatic protective effect compared to non-pregnant women. Pregnancy was associated with a 75 percent reduction in maternal mortality rates compared to that of the nonpregnant women (0.8% vs 3.5%; OR 0.24, 95% CI 0.12–0.48, $p < 0.0001$).[20] Moreover, many suspect that the statement was falsified by a variety of manipulative tactics including the assignment of actual vaccine-induced deaths.

In the same article, Grant writes that Mississippi was seeing a "devastating two-fold increase in the rate of fetal deaths during Covid-19." The blatant sin of omission here is her failure to say what caused these fetal deaths, including whether or not the mothers of the dead babies were vaccinated or unvaccinated. Moreover, as noted above, people were being falsely labelled as unvaccinated for two weeks post-vaccine because, importantly, about 80 percent of Covid vaccine deaths occurred within fourteen days post-vaccination. This proves Grant's figures are absolutely false. She leaves the reader to assume they were all unvaccinated, which is deceptive. Grant also quotes Dobbs saying, "We do know that COVID is especially problematic and dangerous in pregnant women, but we also know that it can be deadly for the baby in the womb." What's being insinuated here is that unvaccinated pregnant women are a danger to themselves and their babies, even after Dobbs had said earlier in the article that the babies of unvaccinated mothers who succumbed to Covid had survived.

Grant's declarative sentences sound true and authoritative but, conversely, they're manipulative to the point of meeting the definition of misinformation.

So, that's how it's done, dear reader.

These organizations and their sources create and disseminate misinformation to cover up the truth while vilifying those who are speaking truth—accusing them of disseminating misinformation. The scale on which they have done this can only be described as demonic.

Toward the end of the article, Grant writes: "The most up-to-date research shows that there is no increased risk of miscarriage or spontaneous abortion after Covid-19 vaccination." She sells her statement's credibility by citing a June 17, 2021, article in the *New England Journal of Medicine*, by Dr. Tom T. Shimabukuro et al, "Preliminary Findings of mRNA Covid-19 Vaccine Safety in Pregnant Persons." This article is discussed elsewhere, but suffice to state clearly that the twenty-one authors, all government employees, deceptively shifted seven hundred patients receiving the vaccine in the third trimester putting them in the denominator of the first trimester, thus falsely changing the miscarriage rate from 82 percent to 12.6 percent.[21,22,23]

NEJM June 17, 2021 Vol 384 No 24, on page 2276: Shimabukuro et al shifted the denominators to dilute the spontaneous abortion rate of 82% (104/127) to 12.6% (104/827). 700 patients receiving the vaccine in the third trimester is gross misrepresentation because a spontaneous abortion is defined as a pregnancy loss before 20 weeks gestation. The 82% spontaneous abortion rate in this *NEJM* report is on par with the "abortion pill" RU486 (Mifepristone). Is this complete ignorance or willful deception?

Finally, the Shimabukuro *NEJM* article's "Funding and Disclosures" section brazenly mentions that all the authors of the study are US government employees or government contractors with no "material conflicts of interest," and that "the findings and conclusions in this article are those of the authors and do not necessarily represent the official position of the Centers for Disease Control and Prevention (CDC) or the Food and Drug Administration." So, if lead author, Dr. Shimabukuro, gets his paycheck from the CDC to serve as director of their Immunization Safety Office [ISO] for the CDC, is that not a direct conflict of interest and a material conflict? They're paying him as director of Immunization Safety, when the CDC itself is a purveyor of misinformation about how safe the vaccines are. Shimabukuro's "findings" certainly line up with the CDC's propaganda aims.[24,25]

These people have no shame.

The September 27, 2021, ABOG Misinformation Statement refers to yet another sin-of-omission; no-proper-context article. As detailed above in the ABOG threat, the ABOG hyperlink "the CDC has reported," does not take you to the CDC website as expected, but to a September 9, 2021, CNN article. This CNN article quotes Dr. Dana Meaney Delman. "We now see increased numbers of pregnant people admitted to the ICU in July and August," she says, adding that the trend continued into September. "The deaths reported in August is the highest number of deaths reported in any month since the start of the pandemic. About 97 percent of the pregnant people treated in the hospital for Covid-19 have been unvaccinated."[26] Again citing the Pineles study above, this statement in the CNN article is false; pregnancy is associated with a 75 percent reduction in maternal mortality compared to non-pregnant women.

Pregnant "people"? Another sign of a world gone mad.

Nevertheless, the same questions apply here: How many among those unvaccinated pregnant women comprising the 97 percent went to the hospital without being offered or knowing about early Covid treatment protocols? How many actually received the vaccine, yet were categorized as unvaccinated? How many survived? How many actually died? What happened to their babies? Without answers to these contextual questions, Delman's information is misleading and virtually meaningless.

Delman's last quote in the article is, "We know Covid-19 vaccines are safe and effective. If you are pregnant, postpartum, breastfeeding, trying to get pregnant now or might become pregnant in the future, please get vaccinated."

# The Threat

This line is literally nauseating to any Ob-Gyn who has experienced what I have. The ABOG September 27, 2021, Misinformation Statement is extremely poorly referenced, with zero credible safety data and pathetic references. The threats represent fascist repression that has no resemblance to science.

My big question is: Why doesn't ABOG and the other organizations just come out and give examples of what they deem to be Covid misinformation? Is it because they know that their examples would not hold up to real scientific scrutiny? They have already been caught using fraudulent studies to make their case.

But is it also because they know they'd face an avalanche of legal trouble if they were exposed engaging in persistently putting out lies to the public that have killed and maimed so many?

Little did they know, they were yet to hear from me. I simply would not let this stand.

## Chapter Eleven

# My Response: An Open Letter from James A. Thorp, MD, to ABOG

*This is a perfect storm that will eclipse the DES and thalidomide disasters and make them look like a sunny day on the beach.*
—James A. Thorp, MD, January 12, 2022, letter to ABOG, final paragraph

### Not Mincing Words

The appalling "September 27, ABOG Misinformation Statement" set me off like a stick of dynamite. As a former officer in the United States Air Force, along with my oath to uphold the US Constitution, I would never tolerate this violation of my First Amendment rights, nor would I tolerate illegitimate threats and impingements on my Hippocratic Oath to my physician-patient relationship. It was literally the clearest transition from academia to fascism that I have ever witnessed in my life. I immediately began exposing ABOG and made up some rather impressive "ABOG-like logos" inserting the communist red-hammer and sickle image and dispersed them on various social media platforms. They were, perhaps not surprising, not received well. Executive director of ABOG, George Wendel, MD, did not appreciate my ABOG logos and threatened me with the measures outlined in his fascistic ABOG Misinformation Statement.

Unfortunately for them, they did not know who they were dealing with: So, to address the threats of George Wendel and ABOG, I composed a letter.

# My Response

Because I had already amassed thousands of hours of research, literature review, and analysis of the CDC/FDA's Vaccine Adverse Event Reporting System (VAERS), I became expert at review and analysis of VAERS. I sent my letter to George Wendel, MD; Susan Ramin, MD; other board members of ABOG; ABOG staff; and ABOG examiners. My online ninety-eight-page letter included 1,019 peer-reviewed medical journal publications—in just twelve months—that documented injuries and deaths after the rollout of the COVID-19 vaccines. Let that sink in: 1,019 peer-reviewed medical journal articles in just twelve months after rollout of this lethal, injurious shot.

As of June 2024, there are now 3,580 such journal articles.[1] Certainly, this was perfectly consistent with the Pfizer 5.3.6 post-market analysis documenting the deadly and injurious effects of their "safe and effective" vaccine.

The letter was immediately published for the entire world to see in perpetuity and is available by simply searching James A. Thorp, MD "Open Letter to ABOG." It is also available on the rodefshalom613.org website.[2]

Most would agree that the ninety-eight-page letter makes the one-page ABOG Misinformation Statement appear to be written by amateurs.

Here's my letter in full (minus the 1,019 references, which are available to view on the Rodefshalom613.org website[3]):

January 12, 2022

Dear Dr. George Wendel and my other esteemed ABOG colleagues,

I appreciate your willingness to dialogue with me. As you know, I have previously expressed my concerns, now shared by a number of my colleagues, regarding (1) the safety of the COVID-19 experimental mRNA and DNA gene therapy injections in pregnancy, and (2) ABOG's disconcerting September 2021 *Statement Regarding Dissemination of COVID-19 Misinformation*, which has blatantly threatened constituents with revocation of their medical license for "providing misinformation about the COVID-19 vaccine." As a 40-plus year member in good standing of ABOG—an organization I have always held in high esteem—I can genuinely say that the intimidating nature of ABOG's September 2021 *Statement Regarding Dissemination of COVID-19 Misinformation* is heretofore unprecedented.

Couched in bullying language, ABOG's *Statement Regarding Dissemination of COVID- 19 Misinformation* begs critically important questions for both constituents and patients alike.

These questions are intrinsic to a physician's ability to treat patients free from conflict of interest over fear of reprisal from ABOG (or others). Even more important, these questions concern matters essential to safeguarding and protecting maternal and fetal health and well-being, and are essential to upholding the physician's oath to do no harm. First, what constitutes that which ABOG deems "misinformation about the COVID-19 vaccine"? Second, by whom and how is such "misinformation about the COVID-19 vaccine" determined?

ABOG's *Statement Regarding Dissemination of COVID-19 Misinformation* fails to provide answers to these questions and also fails to acknowledge the growing body of scientific, peer-reviewed evidence that the experimental mRNA and DNA gene therapy injections are a failed strategy that have killed, injured, and endangered many. ABOG's widely circulated *Statement Regarding Dissemination of COVID-19 Misinformation* has placed patients' health in jeopardy while leaving constituents holding the proverbial bag as their medical license and livelihood hang in the balance. Patient safety is sacrificed as constituents toe the line—forced to choose between pushing experimental gene therapies shown to be dangerous to both mom and fetus or lose their livelihoods.

ABOG's *Statement Regarding Dissemination of COVID-19 Misinformation* turns a blind eye to this ever-growing evidence and dodges these thorny issues by pointing to published statements made by the Federation of State Medical Boards (FSMB) and the American Board of Medical Specialties (ABMS). However, upon further examination, neither the FSMB nor the ABMS provides adequate answers. If anything, the express collaboration of ABOG with the FSMB and ABMS, together with the language of the FSMB's and ABMS's individual statements, raises red flags about potential collusion, bias, and conflicts of interest within the various medical stakeholders, Big Tech, and the media.

### Following ABOG's "Yellow Brick Road": The Federation of State Medical Boards (FSMB)

ABOG's *Statement Regarding Dissemination of COVID-19 Misinformation* first points to the definition provided by the Federation of State Medical Boards (FSMB) for answers.

However, the FSMB fails to provide any clear answers to what constitutes "misinformation about the COVID-19 vaccine," or how such "misinformation" is to be determined, or by whom. Instead of providing answers, the FSMB launches accusations, stating that there has been "a dramatic increase in the dissemination of COVID-19 vaccine misinformation and disinformation by physicians and other health care professionals on social media platforms, online and in the media." The FSMB further notes that physicians "have an ethical and professional responsibility to practice medicine in the best interests of their patients and must share information that is factual, scientifically grounded and *consensus-driven* for the betterment of public health."

But exactly what information is "factual, scientifically grounded and consensus-driven for the betterment of public health?" After all, the stakes for all humanity in getting this right could not be higher. And perhaps more importantly, who is the arbiter of such information? While the FSMB does little to define "COVID-19 vaccine misinformation," it does seem to suggest that somewhere, somehow, a certain "consensus" exists on what narrative *should* prevail, perhaps giving clues as to whose interests this prevailing narrative should serve. Toward that end, the FSMB statement's express reference to "social media platforms" and mainstream "media" is highly disturbing, particularly in light of alleged ties between the mainstream media outlets, social media giants, and Pfizer, suggesting (at a minimum) serious conflicts of interest. Disconcertingly, the FSMB statement seems to give the appearance it has communicated—if not possibly even conspired with—Big Tech and the mainstream media to somehow root out what they collectively deem "misinformation."

### Following ABOG's "Yellow Brock Road": The American Board of Medical Specialties (ABMS)

ABOG's *Statement Regarding Dissemination of COVID-19 Misinformation* next points to the American Board of Medical Specialties (ABMS) for answers regarding what constitutes "misinformation about the COVID-19 vaccine." At first glance, the ABMS statement and definition of misinformation seems just as fuzzy as the statement provided by the Federation of State Medical Boards (FSMB). Up front, however, the ABMS identifies "vaccine hesitancy" as the real culprit—explicitly linking misinformation with vaccine hesitancy. Evidently, any information that does not push mass vaccination with experimental gene therapy on all persons constitutes COVID-19 vaccine misinformation, and could "threaten certification by an ABMS Member Board." Interestingly, the ABMS has attempted to frame its statement on COVID-19 misinformation as one that is *supportive* of medical professionals, titling its press release as follows: *ABMS Issues Statement Supporting Role of Medical Professionals in Preventing COVID-19 Misinformation.* Yet the ABMS' threats contained within the press release do not logically follow from these words.

### Euphemisms, Intimidation, and Gaslighting, Oh My!

Since ABOG itself has declined to define "misinformation about the COVID-19 vaccine," deferring instead to other medical agencies, I will kindly attempt to offer one for ABOG's consideration. Following the lead of Robert J. Kennedy, Jr., the phrase "misinformation about the COVID-19 vaccine" seems to be "a euphemism" for any statement or scientific evidence that differs from the prevailing narrative of stakeholders who most stand to profit from the COVID-19 vaccines. In this case, these stakeholders appear to include Big Tech, Government, the Pharmaceutical Companies, Big Media, and various Corporate and Medical Stakeholders. These are the stakeholders who drive the "consensus" referred to by the FSMB. Perhaps not coincidentally, these are the very stakeholders that are attempting to drive the false narrative that vaccines are safe, effective, and necessary for all persons, including pregnant persons. As it turns out, there is no real definition for "COVID-19 vaccine misinformation"—or real answer to my first question. The phrase "COVID-19 vaccine misinformation" constitutes a euphemism. Euphemisms don't create meaning, they disguise it, and have thus been referred to "the language of evasion, hypocrisy, prudery, and deceit (Holder 2008)."

Gaslighting has been described as "an insidious form of manipulation and psychological control" where victims are "deliberately and systematically fed false information that leads them to question what they know to be true." Gaslighting occurs when "an abuser tries to control a victim by twisting their sense of reality." The abuser or bully misleads their target, creating a false narrative and making them question their judgments, reality, and perception. When the victim calls out the gaslighting, the abuse will frequently try to discredit their victim. When dealing with someone who is gaslighting, it is advised to pay close attention to what the abuser actually does, instead of the words they use.

ABOG's *Statement Regarding Dissemination of COVID-19 Misinformation* is nothing short of gaslighting. ABOG's statement professes to encourage constituents to "practice evidence-based medicine based on facts and scientific data." ABOG's statement also adopts the FSMB's position that its physicians "have an ethical and professional responsibility to practice medicine in the best interests of their patients and must share information that is factual, scientifically grounded and consensus-driven for the betterment of public health." ABOG has expressed a concern for protecting patients from harm. While ABOG's words express concerns about patient safety and stress the importance of scientific data, ABOG turns a blind eye to applying the best available scientific evidence and protecting patients from the dangers of experimental vaccines, all while threatening the medical license of constituents who challenge the prevailing narrative. The foregoing is a classic example of gaslighting. Despite its words, ABOG's threatening conduct tells a different story.

**Exposing the Wizard: "I'm really a very good man; but I'm a very bad Wizard, I must admit"**
ABOG is not alone. Gaslighting has never before occurred as widely as it has during the COVID pandemic, with the pushing of experimental gene injections as the only effective, safe, and necessary option for all persons, even pregnant persons. ABOG, you have a golden opportunity to reverse course, taking a stance that is factual and scientifically grounded, and true to your words purporting to protect patients from harm—by retracting and revising your *Statement Regarding Dissemination of COVID-19 Misinformation*. Will you have the courage to do it? As an organization that professes to care about patients' interests and safety, you may want to peruse the many testimonies of the vaccine injured on Real, Not Rare. If you do, you will find the stories of many who have suffered serious, life-altering adverse effects from the experimental injections, almost all of which share remarkably similar characteristics and symptoms. These vaccine injuries appear to be vastly under-reported. Those who have the courage to come forward are frequently called crazy and accused of mental illness, thus enduring gaslighting by doctors, who often refuse to believe them. ABOG, instead of hiding behind euphemisms, false narratives, and other medical organization's euphemistic verbiage, you can choose to take a bold stand for your patients and constituents and lead the way to exposing what is really going on. Will you do it? The EU in recent days has taken a bold step in this direction—warning that boosters risk adverse effects to the immune system and may not be warranted. And a top Israel immunologist has recently followed suit, calling on its leaders at the Israeli Ministry of Heath to admit that the mass vaccination campaign has failed in Israel.

There is an undeniable and growing body of peer-reviewed scientific evidence that these experimental gene therapy injections are unsafe and dangerous to both mothers, fetuses, newborns and infants. Indeed, since the publication and dissemination of ABOG's *Statement Regarding Dissemination of COVID-19 Misinformation*, the Johnson & Johnson injection is no longer recommended for use after life-threatening blood clots and deaths have been linked to the injection. This growing body of evidence credibly and scientifically calls into question the efficacy of these experimental gene therapeutic injections. As unprecedented numbers of new infections now make painfully clear, the experimental injections are proving to be wholly ineffective at preventing infection of the Omicron strain, the current dominant strain in the US. Multiple recent studies indicate that the vaccinated are more likely to be infected with Omicron than the unvaccinated. For example, numbers in a recent study from Denmark now show persons who received the experimental injections are up to 8 times more likely to develop Omicron that those persons who did not. Multiple independent studies indicate that the more one vaccinates, the more one becomes susceptible to COVID-19 infection. Recent studies also suggest that the COVID-19 gene therapy injections cause more COVID cases per million and more deaths per million associated with COVID. Studies which show the experimental injections to be neither safe nor effective, but outright dangerous, are almost too numerous to count. Patient's own brave testimonies on such sites as on Real, Not Rare, are heartbreaking.

Continuing to require your constituents to push experimental COVID-19 gene therapy injections on patients in light of mounting evidence that they are neither safe nor effective is ignoring science and placing patients in grave danger. ABOG's pushing the narrative that the experimental injections are safe and effective in the face of such evidence amounts to an egregious false misrepresentation and an intentional failure to disclose the truth to patients.

If ABOG truly cared about encouraging the practice of "evidence-based medicine based on facts and scientific data" and "acting in your patients' bests interests"—as it claims in its *Statement Regarding Dissemination of COVID-19 Misinformation*—then it would retract its statement and the threats issued to its constituents. ABOG would welcome and consider independent, unbiased scientific data which seek to challenge the safety and efficacy of the gene therapy injections. ABOG would be willing to challenge the prevailing stakeholder consensus/orthodoxy/narrative that the experimental injections are safe, effective, and necessary. For ABOG to do less than the foregoing, while at the same time professing to care about science and patient safety, is nothing short of gaslighting.

### Changing Course?

It is my desire to work with you, not against you, and assist to reverse the dangerous course that ABOG has taken. I believe that ABOG's current course is headed for extreme and unparalleled disaster with untold human lives at stake. I would offer my services to you on a *pro bono* basis, and I think we could begin to work through these pressing issues that have threatened the

# My Response

care of women of reproductive age, pregnant women, and their newborns and infants.

Two main issues require your immediate attention that would be best addressed in a formal statement to your constituents. First, the *Statement Regarding Dissemination of COVID- 19 Misinformation* published on the ABOG website and circulated to ABOG's 22,000 plus specialists and sub-specialists in September 2021 needs to be formally retracted. Second, ABOG needs to immediately specifically recommend *against* the vaccine in pregnancy until there are long-term safety data in the offspring. Experimental gene therapy in pregnancy is extremely radical and without historical precedent absent safety data. It is completely unnecessary as there are much safer alternatives for prevention of all viral illnesses. To push the experimental injections violates our Hippocratic oath of informed consent and *Primum Non Nocere*.

I know most of you. I feel confident that it is not your intent to issue unethical threats, gaslight, and act in contravention of mounting medical and scientific evidence casting serious doubt on the safety and efficacy of the experimental gene therapy injections. I believe it is not your desire to intentionally mislead and harm patients and fetuses, or destroy informed consent and the sanctity of the doctor-patient relationship. As a long-standing supporter of this distinguished organization, I find it difficult to believe that you personally created the language contained in your *Statement Regarding Dissemination of COVID-19 Misinformation*. Perhaps not coincidentally, this same language appeared and was published almost simultaneously by FSMB.org, ABMS.org, AANC.org, multiple other ABMS boards, the AMA, SMFM, ACOG, governmental and private organizations with ties to the pharmaceutical industry, the pharmaceutical industry itself, "Big Tech" companies, the mainstream media, multiple medical journals, insurance companies, and many others who have a financial or other stake in pushing the experimental gene therapies. For lack of a better term, I have collectively labelled the forgoing as a CARTEL, as this best describes the blatant conflicts of interest and collusion which lies at the heart of this language. If ABOG is truly "acting in patients' best interests,"—as it proclaims to do—it should be fully committed to the health of its pregnant patients, not serving the interests of the pharmaceutical industry or any other organization. It was and is your responsibility to resist and oppose this inappropriate language that was most likely pressed upon you by ABMS, FSMB and others.

Sir Karl Popper stated that academicians and science should always be open to divergent opinions, and the scientific method includes discussion of opposing minority opinions and views. Science progresses *only* by refutation.

In fact, as history has demonstrated, absent the foregoing, there can be no true science at all. Galileo was persecuted because of his minority opinion that our solar system was heliocentric, not geocentric. Likewise, in the mid 1800's Ignaz Phillip Semmelweis opposed the mainstream narrative. He proved that the 50% maternal mortality at Vienna Lying in Hospital could be decreased simply by hand washing. However, he was mocked, derided, and persecuted by the 'ABOG-like authorities' of his time. Similarly, those of us who expressed concern and opposed the liberal use of opioids in the 1990's (including myself) in our patients were reprimanded; this greatly contributed to the major opioid crisis our country experienced, thus fulfilling the law of unintended consequences. These examples underscore what ABOG is currently doing to their constituents, and it is reaping unparalleled disaster.

ABOG's narrative is *not* evidence based. It was incumbent upon ABOG to have demanded safety studies WITH LONG TERM OUTCOMES *prior* to issuing their threatening language. It is not incumbent upon the vulnerable and innocent to prove that they have been harmed. ABOG is on the wrong side of truth and is forcing a false narrative that will be ultimately responsible for killing and injuring many more than they already have. ABOG has implemented a dangerous retrovirus gene therapy in pregnancy and in women of reproductive age with zero credible studies conducted on whether it is safe.

ABOG is obviously aware that there are multiple OB/GYN and Maternal Fetal Medicine 'experts' who have no credibility and no data as to safety, yet they are pushing this dangerous COVID-19 retrovirus gene therapy in pregnancy and in women of child-bearing age all over the US and the world. Any attempt to engage in informed consent and/or contradict the ABOG narrative comes with the very real risk of damaged professional careers and irrevocably lost livelihoods. Indeed, a national townhall meeting of two maternal fetal medicine physicians took place recently in Indianapolis with the sole purpose of pushing the experimental gene therapy on pregnant women across the nation. Both young MFM physicians were old enough to be my daughters and had a combined clinical experience/publications of only a small fraction of mine. Yet neither I nor any of my colleagues opposing this dangerous recommendation were allowed to participate and balance their outrageously absurd and dangerous presentations. There are so many doing the exact same thing because of professional threats which have been placed on the careers of doctors who oppose the false "safe and effective" narrative. This too is ABOG's fault.

The threats that you have circulated to all your specialist and subspecialists have resulted in the pushing of the experimental gene therapy injections in

# My Response

*all* women of reproductive age and pregnant women, with extremely serious consequences. There are multiple independent sources all over the world that have observed significant increases in miscarriage (spontaneous abortion), fetal death, fetal malformation, severe placental inflammation, severe IUGR, neonatal demise, infant demise, permanent newborn/infant/child chronic autoimmune diseases, permanent immune deficiency syndromes, chronic permanent CNS diseases and chronic cognitive impairment, seizure disorders and the unleashing of neonatal / infant cancers and opportunistic infections, and many other disastrous consequences. Please understand and recognize it is not incumbent upon me to prove this to ABOG or anyone else; rather, it was incumbent upon those recommending it to show safety data *before* pushing this extremely dangerous experimental gene therapy in all pregnant women.

Animal studies clearly demonstrate that the lipid nanoparticles (LNPs) with their mRNA cargo pass through all God made barriers, including the blood brain barriers, the placental barrier, and the fetal blood brain barriers. It is also known that there is a very significant concentration of the LNPs in the maternal, fetal, and newborn ovaries. As you well know, a female fetus is born with all of its gametes (about 1 million ovum) in the ovaries and each is exposed to these potentially poisonous NLP's. It is now widely known and understood that the 'vaccine,' which is actually an experimental gene therapy, works by inducing inflammation. Yet, *inflammation in the developing embryo and fetus is a hallmark for permanent damage, malformation, death, placental insufficiency, and potentially life long chronic diseases in the offspring including severe immunological disturbances, disruption of the TOL7 and TOL8 receptors on cell membranes.* The disruption of the TOL7 and TOL8 receptors are responsible for immune surveillance and the suppression of cancers and opportunistic infections in the body including herpes, zoster, CMV, HPV, TB, toxoplasmosis and many others. Dr. Ryan Cole MD a highly acclaimed pathologist has noted a striking increase in newborn and infant cancers that are extremely rare. There are unprecedented numbers of stillbirth in the US, Canada, Scotland, Europe and many other locations. Scott Davison the CEO of OneAmerica insurance company has noted all-cause death rates are up 40% in ages 18-64 years; this is unprecedented and certainly not attributable to just COVID-19 but also to the experimental gene therapy that you have pushed. He notes that even a 10% rate of increase is catastrophic for the insurance industry. Life insurance companies are warning that there are nearly 100,000 excess deaths per month happening in all age groups in the USA and cannot be attributed to COVID-19.

It is also important for all of ABOG to recognize that they cannot defer this blame to ABMS, FSMB, AMA, ACOG, SMFM, AMA or any other medical organization. ABOG is responsible for all these consequences in pregnant women and women of childbearing age, since this is ABOG's jurisdiction. As you know, ABMS, FSMB, SMFM, ACOG, SMFM and AMA have *no* authoritative action; their recommendations and threats are 'teethless'. This is ABOG's jurisdiction, not the jurisdiction of SMFM, ACOG, AMA or any other of the organizations because they have no authority to threaten their constituents like ABOG has done to me personally and all the other ABOG specialists and subspecialists. As I served on the SMFM BOD for several years, I understand SMFM. They do not discipline their constituents. They are a social organization which provides education and specifically does *not* have the authority to sanction or call out constituents. All of the blame for the mass casualties can only lie with ABOG's decision to push the experimental gene therapy in reproductive age women.

As to safety, the Vaccine Adverse Event Reporting System (VAERS) alone has signaled that the experimental mRNA gene therapy is dangerous. While perhaps not perfect, this data simply cannot be ignored, denied, or derided by ABOG or any other organizations or treating physicians. VAERS is a statutorily created safety surveillance tool created as an outgrowth of the National Childhood Vaccine Injury Act of 1986 (the "Act"). *See* 42 USC § 300aa-1 *et seq.* Administered by the CDC and FDA, the creation of VAERS was essentially a *quid pro quo* arrangement for the blanket sovereign immunity given to the pharmaceutical companies for vaccine research and development. Under the Act, pharmaceutical companies are given full immunity from tort litigation arising out of injuries and damages relating to vaccines. However, physicians and organizations like ABOG are *not* similarly situated. Rather, under the Act, as the eyes and ears of patient care, physicians have an affirmative duty to report to adverse events, and VAERS has been deemed "the front line" of vaccine safety. As the 2000 Committee on Government Reform aptly noted:

> *The Act does require that physicians report*—directly to VAERS or to the manufacturer—certain categories of serious outcomes defined for regulatory purposes as an event resulting in death, life- threatening illness, hospitalization, prolongation of existing hospitalization, or permanent disability.
>
> ***VAERS is intended to serve as the "front line" of vaccine safety***, since this type of national reporting system can rapidly document possible effects and

# My Response

*generate early warning signals that can then be more rigorously investigated in focused studies.* **VAERS is considered especially valuable in assessing the safety of newly marketed vaccines.**

Physicians serve a critical reporting role in patient safety by virtue of this statutorily created reporting system. Failing to take seriously VAERS, which is the "front line" of vaccine safety, turning a blind eye to VAERS signal data, and/or pushing a false narrative in the face of VAERS evidence to the contrary flies in the face of those who took the oath, "first, do no harm." It smacks of serving the pharmaceutical interests over the patients' interests. Pushing the false narrative that the experimental COVID-19 mRNA gene therapies are safe and effective in pregnancy, when VAERS data clearly signals otherwise, constitutes a great and tragic ethical and moral failing by those tasked with being on the front lines of patient care and safety. It also falsely misrepresents the true data, and is not faithful to the stated mission of ABOG. It fails to disclose to unsuspecting patients that the safety of the experimental COVID-19 injections has been called into question, and thus both misrepresents and hides information that is material to the patients' decision. It is not unreasonable to ponder whether such egregious conduct and omissions could expose, physicians or physician entities such as ABOG, SMFM or ACOG to legal and/or even criminal liability at some point in the future. Unlike the pharmaceutical companies, there is no blanket immunity for ABOG and staff or physicians who are on the front lines of safeguarding the health of moms and fetuses.

### The VAERS Data Has Signaled Warnings that Can No Longer Be Ignored

VAERS has shown that the experimental COVID mRNA gene therapy injections have proven to be harmful by any modern safety standards traditionally applied to other vaccines. The "5/50 rule" has always been a "rule of thumb"; if there are 5 deaths associated with a drug, vaccine or device then a black box warning is issued; if there are 50 deaths the product is immediately removed from the market. Why has ABOG now ignored these long-held safety monitors?

Please see my Power Point slides below from my personal analytics from the CDC/FDA VAERS database via medalerts.org. This was analyzed in conjunction with a Silicon Valley IT expert procurator of this database. In my VAERS analytics I created a six-slide set copied below. I compared 4 bars on the horizontal axis representing the COVID-19 vaccine, all other vaccines in VAERS other than COVID-19 vaccine, the influenza vaccine, and the pertussis vaccine. I purposely reviewed the influenza vaccines and the pertussis vaccines because I have personally reviewed the extensive data with these two vaccines in pregnancy and I believe they might provide a risk / benefit that is potentially favorable for the mother/fetus/newborn/infant. The slides are stratified by vaccine associated deaths, fetal malformations, and pregnancy loss. I purposely avoided spontaneous abortions and fetal deaths individually but included them together in pregnancy loss; I do not believe the VAERS database enterers had our expertise to differentiate between these diagnoses. The COVID-19 deaths, fetal malformations, and pregnancy loss are stunning when compared to all other vaccines in the VAERS registry combined, the influenza vaccines, and the pertussis vaccines. Regardless of how you criticize my analytics they are extremely robust, complete, broad and irrefutable. You are morally, ethically, and legally obligated to follow this verified safety signal.

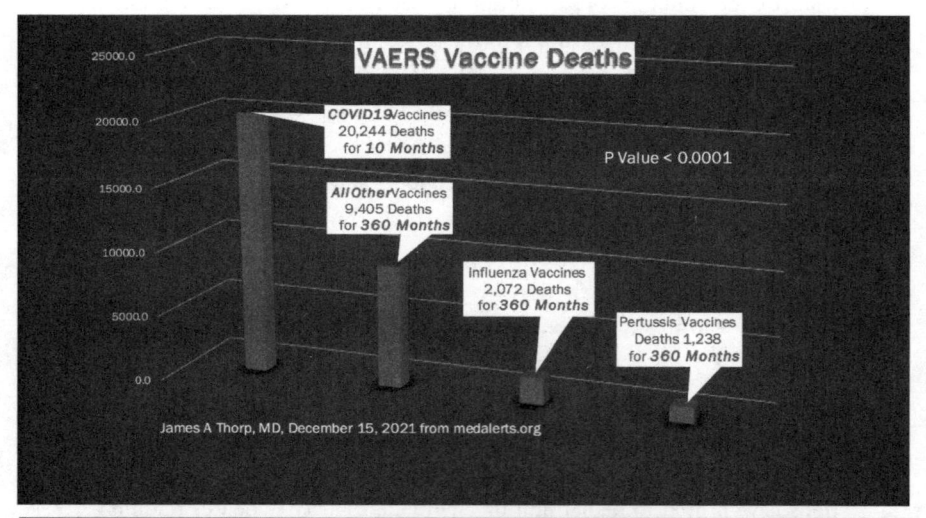

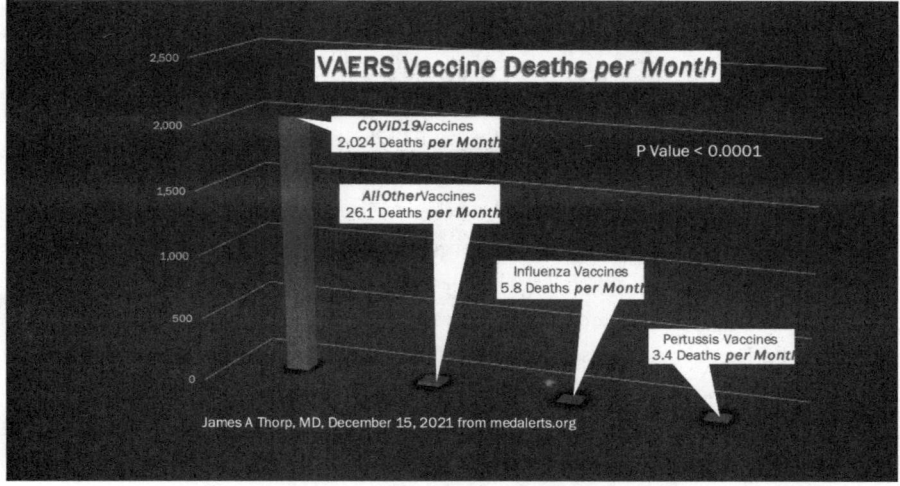

### VAERS Vaccine Deaths

- COVID19 Vaccines: 20,244 Deaths for **10 Months**
- All Other Vaccines: 9,405 Deaths for **360 Months**
- Influenza Vaccines: 2,072 Deaths for **360 Months**
- Pertussis Vaccines: 1,238 Deaths for **360 Months**

P Value < 0.0001

James A Thorp, MD, December 15, 2021 from medalerts.org

### VAERS Vaccine Deaths per Month

- COVID19 Vaccines: 2,024 Deaths *per Month*
- All Other Vaccines: 26.1 Deaths *per Month*
- Influenza Vaccines: 5.8 Deaths *per Month*
- Pertussis Vaccines: 3.4 Deaths *per Month*

P Value < 0.0001

James A Thorp, MD, December 15, 2021 from medalerts.org

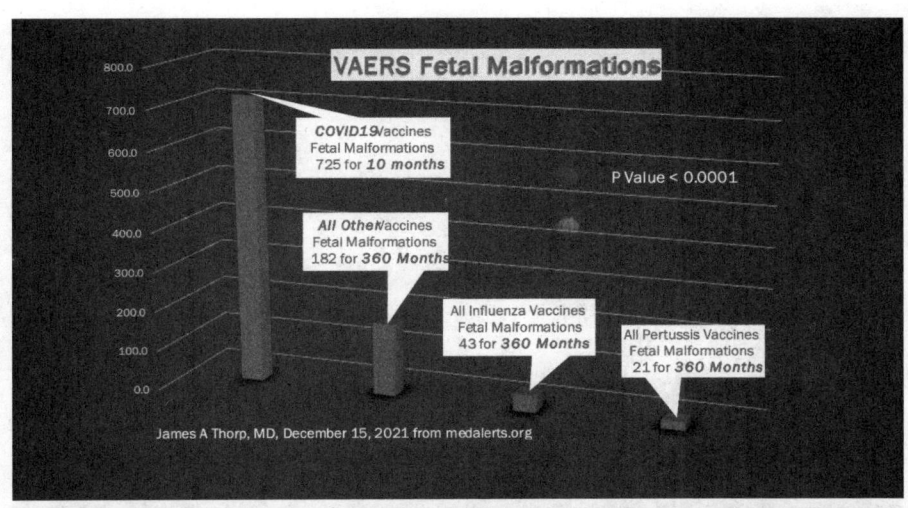

# My Response

## VAERS Fetal Malformations

- COVID19 Vaccines Fetal Malformations 725 for **10 months**
- All Other Vaccines Fetal Malformations 182 for **360 Months**
- All Influenza Vaccines Fetal Malformations 43 for **360 Months**
- All Pertussis Vaccines Fetal Malformations 21 for **360 Months**

P Value < 0.0001

James A Thorp, MD, December 15, 2021 from medalerts.org

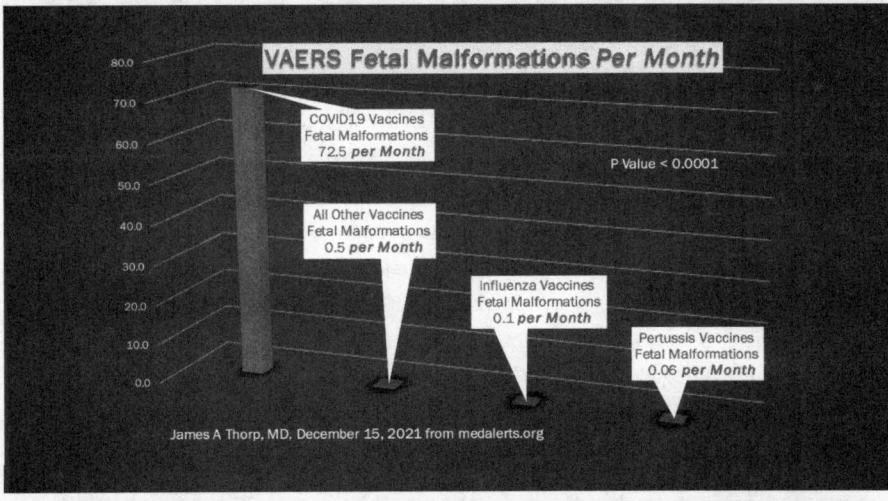

## VAERS Fetal Malformations Per Month

- COVID19 Vaccines Fetal Malformations 72.5 **per Month**
- All Other Vaccines Fetal Malformations 0.5 **per Month**
- Influenza Vaccines Fetal Malformations 0.1 **per Month**
- Pertussis Vaccines Fetal Malformations 0.06 **per Month**

P Value < 0.0001

James A Thorp, MD, December 15, 2021 from medalerts.org

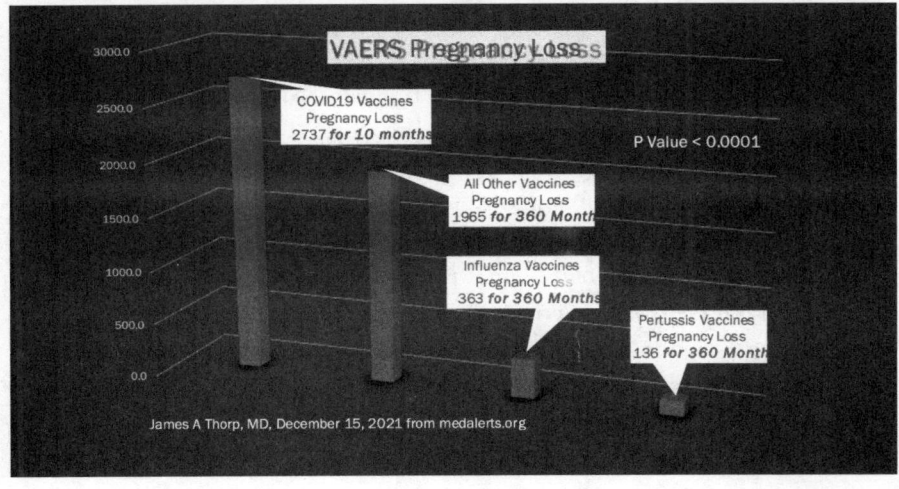

## VAERS Pregnancy Loss

- COVID19 Vaccines Pregnancy Loss 2737 **for 10 months**
- All Other Vaccines Pregnancy Loss 1965 **for 360 Month**
- Influenza Vaccines Pregnancy Loss 363 **for 360 Months**
- Pertussis Vaccines Pregnancy Loss 136 **for 360 Month**

P Value < 0.0001

James A Thorp, MD, December 15, 2021 from medalerts.org

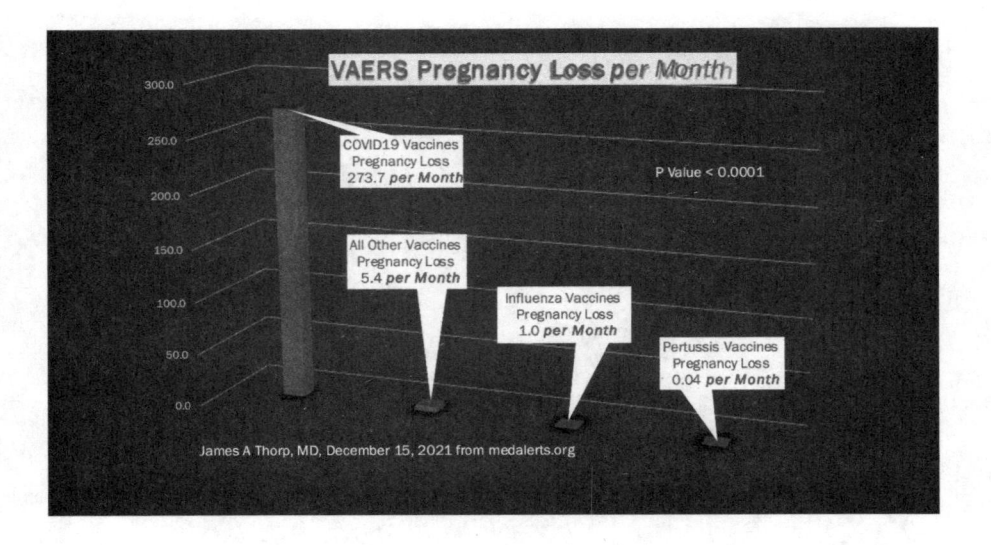

**VAERS Pregnancy Loss per Month**

- COVID19 Vaccines Pregnancy Loss 273.7 *per Month*
- All Other Vaccines Pregnancy Loss 5.4 *per Month*
- Influenza Vaccines Pregnancy Loss 1.0 *per Month*
- Pertussis Vaccines Pregnancy Loss 0.04 *per Month*
- P Value < 0.0001

James A Thorp, MD, December 15, 2021 from medalerts.org

There are multiple reports from all over the world reporting an increase in fetal deaths, neonatal deaths, and infant deaths. The Scottish data documents a 25% increase among infant deaths.

There are NO credible safety data that you have in pregnancy. ZERO. The NEJM article has absolutely nothing safe about it; the VSAFE procurement of data is totally irrelevant, unreliable and laughable. The longitudinal study in the very poorly done study lasted less than 8 weeks; last I heard, pregnancy lasted for 40 weeks. This is complete nonsense. There was ZERO newborn follow up. The fact that ABOG would push this vaccine in pregnancy is an abomination and will be the greatest disaster in the history of obstetrics. The article in the *British Journal of Medicine* in November 2021 by Paul D. Thacker whistleblower exposes the major flaws in the data, fraudulent data, manipulation of data and unblinding of the 'blinded data' and the fraudulent relationship between Pfizer and the NEJM.

The Editor in Chief of the NEJM, Dr. Eric Rubin, had the unethical audacity to vote in favor of the 'vaccine' in children ages 5-11 at the FDA advisory committee, despite being on public record as stating the following: **"We're never gonna learn about how safe the vaccine is until we start giving it."** This is absolutely the antithesis of the scientific method. At least Dr. Rubin was honest in this case. In the case of pregnant women, he simply published and pushed the vaccine in pregnancy with false and worthless data stamped with the seal of the NEJM which most revere as a false god.

The mainstream medical journals have become increasingly corrupted in the last few decades with unethical entanglements with industry and politicians. How is it that 14 months ago the LANCET published a manuscript with the lead author from Harvard that consisted of completely fraudulent data; it was not manipulated data, it was **completely falsified** for the specific political purpose of doing a 'hatchet job' on hydroxychloroquine. The LANCET got caught with their 'pants down' and was forced to retract the article, yet the Harvard professor was not punished. And ABOG threatens to destroy their constituents if they spread 'misinformation'? Why did the LANCET publish this completely fraudulent article? It is pure and simple. They wanted to eliminate an extremely safe and effective drug for the early treatment of COVID-19 so that the pharmaceutical complex could make trillions of dollars on the fraudulent experimental gene therapy. There are thousands of physicians like myself that have been treating COVID-19 very early, effectively and safely with a variety of vitamins, supplements, ivermectin, hydroxychloroquine, azithromycin, doxycycline steroids, and other safe medications. Had the CARTEL not mocked, derided, threatened, censored and persecuted those of us who have been using early outpatient treatment then well over 80% of the deaths and hospitalizations in the US never would have occurred. Hydroxychloroquine has been used for 85 years with billions of doses dispensed and has a higher safety profile than aspirin or acetaminophen. Obstetricians, Maternal Fetal Medicine physicians have used hydroxychloroquine for over 40 years in pregnancy without any adverse effects or teratogenicity noted. Hydroxychloroquine in contrast to the experimental gene therapy is extremely safe and also efficacious in therapy during pregnancy and prophylaxis in pregnancy (400 mg PO once weekly).

Dr. Bhakdi and Dr. Arne Burkhardt explain the basic immunology and unequivocally document the unexpected deaths in Germany are caused by the experimental gene therapy; not by any other causes. As Dr. Bhakdi and Burkhardt explain, death is caused by the autoimmune "self-attack" of T killer lymphocytes in all organs systems throughout the body. He states emphatically that all the gene-based vaccines are killing the young and the old.

This corruption makes the recent high-profile fraud trial involving Stanford dropout Elizabeth Holmes—who bilked hundreds and millions of dollars out of investors and hoodwinked the likes of James Mattis and Robert Murdoch in connection with lies she weaved when her medical tech startup Theranos failed—look like child's play. This potential fraud is exponentially worse than the Theranos disaster and amounts to a medical disaster of unparalleled proportions. The Omicron variant exposes the major flaws in any of the COVID-19 vaccines.

The vaccine does NOT have positive efficacy; it actually has **NEGATIVE EFFICACY.** It increases viral loads, transmissibility and increases the risk to those vaccinated. This is a disease of the vaccinated, NOT the unvaccinated. The most vaccinated countries in the world have the highest case rates of COVID-19. The most vaccinated states in the USA have the highest rates of COVID-19. How is it that the entire USA NAVY ship the USS Milwaukee had 100% of its crew vaccinated out in the middle of the ocean and it is rendered completely disabled because of a massive COVID-19 break out? How is this possible? The false narratives have fallen apart right in front of everyone's eyes.

### The Overwhelming Evidence

*In just 12 months from deployment of the COVID 19 vaccine I have reviewed approximately 1,019 references noted below.* I would encourage ALL the ABOG staff and examiners to review all these studies, as I have done. After you read all these 1,019 peer- reviewed publications listed below in the references you can come to no other conclusion except that the COVID-19 experimental gene therapy injections are highly morbid and mortal in women of reproductive ages, pregnant women, their offspring and children that will be given this experimental gene therapy injection. The dangers of the COVID-19 experimental gene therapy are irrefutable and ABOG must retract their September 2021 threats and make a strong statement regarding the risks of this experimental gene therapy injection in women of reproductive age and in pregnancy. Please do your due diligence and read these references, as I have done.

I will personally debate anyone in ABOG in a public forum or recorded on Zoom for public consumption. As a matter of fact, I have offered to debate anyone in the world on this topic for a year and I have had no one challenge me, only *ad hominem* attacks. Steve Kirsch has offered $1,000,000 US currency to anyone in the world that will debate him. Why doesn't ABOG take him up and pocket $1 million? Here is his phone number: 650-279-1008. He welcomes your call.

This is a perfect storm that will eclipse the disasters of DES and thalidomide and make them look like a sunny day on the beach.

I look forward to working with you and assist you in reversing this madness. Thank you very much for your time and consideration with this matter.

All The Best,
James A Thorp, MD Board Certified ObGyn
Board Certified Maternal Fetal Medicine

# My Response

On January 31, 2024, the statements I made in the above letter to ABOG written two years earlier were proven true.

The COVID-19 vaccine lipid nanoparticles (LNPs) containing the dangerous mRNA crosses the placenta into the fetal blood and becomes bioactive with production of spike protein in the placenta and in the lining of the uterus (decidua) next to the preborn baby. The LNPs are likely distributed throughout the preborn baby including brain/ovaries/testes/thymus and other organs well.

After I wrote this letter in January 2022, Alden and colleagues demonstrated that in human liver cells in vitro the COVID-19 vaccine mRNA is encoded into the human DNA (genome) by an enzyme called reverse transcriptase.[4] Later in 2022 and again in 2023, two separate studies documented that women who received COVID-19 vaccination excreted the intact COVID-19 vaccine mRNA into their breast milk, potentially "vaccinating" their breastfed infant.[5,6]

## Chapter Twelve

# Vaccine Masquerade: The Rise of the Overnight Vaccine Cult and the Attack on Reproduction

*The world, that understandable and lawful world, was slipping away.*

—William Golding, *Lord of the Flies*

In the summer of 2020, I was a wreck; I felt an indescribable sense of dread, as the Covid vaccine propaganda came to envelop and overtake our entire society. When they came up with the twisted name "Operation Warp Speed," I knew there was no prayer they would backtrack, reconsider, or show any form of sobriety or caution. I was correct in my dread. Then New York City Mayor, Andrew Cuomo, appeared on TV screens as a kind of brutish sex symbol, screaming red-faced that he needed more ventilators, and only wanted to see "Shots into arms. Shots into arms."

It was violent, fanatical, and medically as dangerous as a Jim Jones rally in the jungle. People had no grip on their own senses; they were just crazed to get those shots into their own arms, as they became hypnotized to believe this was the amulet against impending death from Covid. Very few people, if any, were at any significant risk. Almost everyone was treatable—Covid was treatable. Curable. Not a big deal.

Wait, I thought, how could this be? Doesn't a gene product, what they were calling a vaccine, take much longer to develop? Dr. Peter McCullough testified in 2022 that he had "served on two dozen data safety monitoring boards for large pharmaceutical and device and in vitro diagnostic studies."[1] He said that a gene therapy has always required at least five years of study before being rolled out to the public.

The agenda felt more like a sudden eruption of a vaccine cult than a "public health measure." It was obvious. They all lied, all day long through our TV sets. It resembled a mass, very ambitious MK-Ultra type mind control operation. The spate of outrageous lies was continually told and irrational, insane "decisions" were made each day.

Lockdowns. Businesses closed. Schools closed. Children in masks. No visitors to nursing homes. But the biggest, unprecedented lie of all, contradicting millennia of medical experience, and shoved down our throats, was that there was no early treatment available for Covid.

Hydroxychloroquine, which I personally used along with many other therapies, to treat Covid during pregnancy, was villainized. Hydroxychloroquine was removed from global use after the *Lancet* medical journal's May 2020 study (later retracted) concluded it was unsafe. Suddenly, the CDC and FDA yanked decades-long narratives about how safe and effective hydroxychloroquine was from their websites. Luckily, I retrieved them, so I have proof of their unconscionable and deadly about-face. The propaganda that pushed the lies of the "dangers" of hydroxychloroquine, cleared the path to granting the Emergency Use Authorizations (EUA). Without an existing "safe and effective" treatment (hydroxychloroquine), governments could push Covid "vaccines" onto an already traumatized public.

The FDA crept further down their path of deception in September 2021, when it discreetly changed the definitions of "vaccine" and "vaccination." (See below with my bold italics):

**VACCINE:**
    Before September 1, 2021:
        Definition of "*Vaccine*"—A product that stimulates a person's immune system to produce immunity to *a specific disease, protecting the person from that disease.*
    After September 1, 2021:
        Definition of "*Vaccine*"—A preparation that is used to stimulate the body's immune response against *diseases.*[2]

**VACCINATION:**
    Before September 1, 2021
        Definition of "*Vaccination*"—The act of introducing a vaccine into the body *to produce immunity to a specific disease.*

### After September 1, 2021

Definition of "*Vaccination*"—The act of introducing a vaccine into the body to *produce protection from a specific disease*.[3]

These revised definitions arouse suspicions that the CDC was trying to cover up the fact that the "vaccines" were proving to not be 100 percent effective at preventing Covid infection, or 100 percent safe. Kentucky congressman Thomas Massie opined in a tweet that the CDC had "been busy at the Ministry of Truth."[4] In CDC's defense, a spokesperson told *McClatchey News* that the old definitions could have been "interpreted to mean that vaccines were 100 percent effective which has never been the case for any vaccine, so the current definition is more transparent and also describes the ways in which vaccines can be administered."[5]

Pfizer whistleblower Brook Jackson pointed out in a September 2023 tweet that the US military's definition of a vaccine did not change: "The mRNA shots do not contain a single molecule of the Covid-19 virus, and therefore the mRNA shots are not "vaccines" under DODI 6205.02."[6]

To make matters even more disorienting, the German scientist Christian Drosten and his team designed the WHO PCR protocol for the COVID-19 PCR "test" "without having virus material available."[7]

If there was a COVID-19 "virus," then why didn't they use *it* to create the global standard PCR "test" for Covid?

The White House message to the public seven months earlier cleverly avoided the "100 percent effective" issue but used the 100 percent figure to fraudulently inspire confidence about other things. In February 2021, members of the Biden-Harris Transition Covid-19 Advisory Board wrote an opinion piece in *USA Today*: "Take Whatever Covid Vaccine You Can Get. All of Them Stop Death and Hospitalization."[8]

Instead of addressing public concerns about the varying effectiveness rates among the vaccines, the team avoided that and focused on death and hospitalization, making the wildly inaccurate claim that "the vaccines were all 100 percent effective in the vaccine trials in stopping hospitalizations and death." "Waiting for a more effective vaccine," they warned, "is actually the worst thing you can do to lower your risk of getting severely ill and dying of COVID-19." They also wrote that "all seven COVID-19 vaccines that have completed large efficacy trials—Pfizer, Moderna, Johnson & Johnson, Novavax, AstraZeneca, Sputnik V and Sinovac—appear [note the difference between the article's title declaring all the vaccines stop death/hospitalization and this line using the word "appear"] to be 100 percent effective for serious complications. Not one vaccinated person has gotten sick enough to require hospitalization. Not a single vaccinated person has died of Covid 19."

Six weeks after the piece came out, six people died after taking the Johnson & Johnson vaccine and the FDA hit the pause button. Altogether, the vaccine caused "vaccine induced immune thrombotic thrombocytopenia" [VITT] in sixty people. VITT symptoms include blood clots and a low platelet count. In our publication, I reviewed 1,366 peer-reviewed medical journal publications within fifteen months documenting vaccine injury or death after the rollout, and 209 of these reports involved VITT symptoms.[9] Anyone who took the adenovirus Covid-19 vaccines, was at risk for it, but the FDA downplayed that risk by saying it was "rare." Adenoviral vaccines were sold by Vaxzevria (AstraZeneca), Johnson & Johnson/Janssen, and Sputnik V—three of the vaccines Biden's Covid Advisory Board exhorted everyone to take as soon as they could.

Meanwhile, anyone refusing to succumb to the Orwellian diktat demanding that these dangerous, untested gene therapy jabs they called vaccines was met with a wall of vitriol, censorship, bullying, reputation attack, and sometimes de-licensing. In Europe, some opponents of state Covid "science" were criminally charged, and some even served time in prison.

That summer of 2020, I saw all this as a speeding, out-of-control freight train and began to ask myself: If I were in charge, how would I handle this situation? The first thing that came to mind was designing a proper clinical trial to test the safety and efficacy of these gene therapies in vaccine disguise. Because of my statistical background and experience in designing randomized, double-blind, placebo-controlled clinical trials in the 1990s, I knew I could do this.

The first task in designing a prospective clinical trial is to discern its purpose and the main outcome variable. I wanted to determine if these novel genetic treatments were safe and effective in saving lives. The next question—what are the metrics of the outcome, and how many patients will it take?

I did extensive background studies on these therapies and my concern was that this therapy could cause an explosion of chronic diseases, specifically due to chronic inflammation and autoimmune diseases, within ten years after the trial was initiated. I hypothesized that there would be a fivefold increase in death within ten years after the vaccine.

The next metric required to define a proper sample size is what the "effect size" of the death rate between the placebo and experimental group is, and over how long a time period. I postulated that this gene therapy would increase the death rate from 1 in 10,000 to 5 in 10,000, a fivefold increase in death after ten years of treatment. In other words, my ideal design of a randomized, double blind, placebo-controlled trial would not be unblinded for ten years. Many vehemently objected to that. But what they didn't understand, was that I planned an annual interim analysis to be done by a team of real experts, blinded, and separated from all the participants and investigators. This independent interim analysis team would then have the authority to interrupt the study at every annual review if there were statistically conclusive results, positive or negative.

The standard acceptable "power analysis" would require a sample size of about 70,000 patients, including the number that would drop out and/or be lost to follow up. Thus, 35,000 patients would be in the placebo group and 35,000 in the experimental group. A total of 70,000 patients was needed to provide an 80 percent chance of demonstrating a statistically significant 5-fold increase in death from 1 per 10,000 to 5 and 10,000 in 10 years.

While I was very concerned about the "vaccines," I never anticipated that it would have been the deadliest drug ever rolled out. I did not expect the immediate die-off that we obviously experienced, with 50 percent of the deaths occurring within forty-eight hours and 80 percent within one to two weeks, according to some experts. Had my proposed study design been implemented, the first annual interim analysis study group would have IMMEDIATELY terminated the clinical trial and pulled the "vaccine" or novel genetic therapy off the market. Just review the stunning CDC/FDA Vaccine Adverse Events Reporting System (VAERS) report and recognize that the under-reporting factor for VAERS according to many experts (Denis Rancourt, Dr. Jessica Rose, Steve Kirsch, and others) may be forty times the numbers in the graph below.

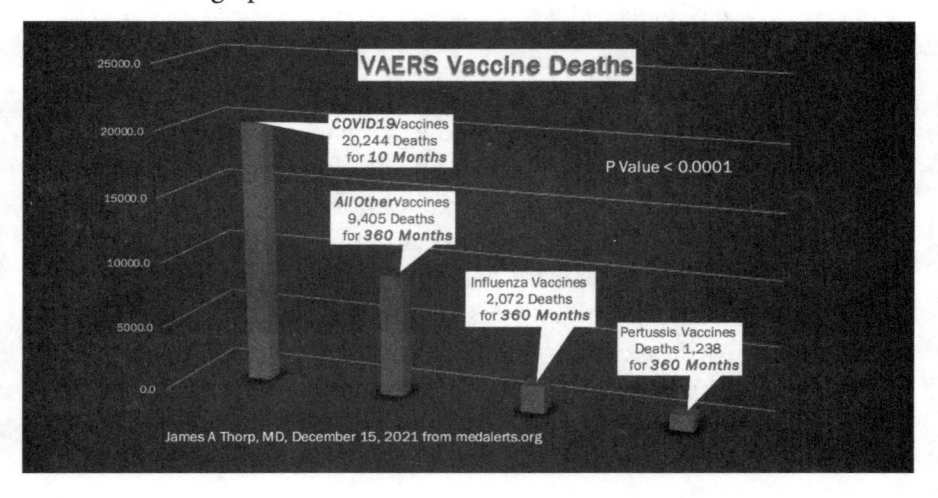

Now, three years after I designed the study, it is difficult for me not to get furious when I see the purposeful manipulation of science from leaders like Dr. Anthony "I Am Science" Fauci and others. The hunch I had after doing extensive background research—before designing the study, that deaths, chronic inflammation, and autoimmune diseases would explode—is coming true.

## Pfizer's Clinical Trial to Evaluate the Safety of the COVID-19 Vaccine in Pregnancy (Phase 2/3) Shows Significant Harms to Babies

On July 14, 2023, Pfizer quietly released the results of its phase 2/3 trial, which was completed a year earlier on July 15, 2022, and had been conducted to evaluate safety of the COVID-19 vaccines received during pregnancy.[10] Pfizer's phase 2/3 trial resorted to all the standard pharmaceutical tactics: 1) Pfizer grossly under-powered the study, using only 315 study participants so as not to achieve a statistical significance (consider that a similar study proposed by Thorp in the summer of 2020 required 70,000 patients!); 2) Pfizer delayed release of the study results for as long as possible, until the product had already done its damage and was presumed to be safe; 3) Pfizer attempted to bury concerning safety data in large amounts of irrelevant information, in an effort to frustrate and obfuscate the readers; and 4) Pfizer hid critical safety data in difficult and unexpected places. This is all standard pharmaceutical operating procedures of deception that have been used for years and are well-described in the book, *Turtles All the Way Down: Vaccine Science and Myth*.[11]

Below are some of the risks which can be found in the Pfizer's phase 2/3 trial:

- Low Apgar Scores (depressed newborns) increased by 100 percent;
- Meconium Aspiration Syndrome (a life-threatening complication) substantially increased;
- Neonatal Jaundice increased by 80 percent;
- Congenital malformations (birth defects) increased by 70 percent;
- Atrial Septal Defect (a hole in the heart) increased by 220 percent;
- Small for Dates (pre-born baby with poor growth) substantially increased;
- Congenital Nevus (vascular malformation on the skin) increased by 200 percent;
- Congenital Anomalies and Developmental Delays at 6 months of life increased by 310 percent.

Would you take the COVID-19 vaccine while pregnant if your Ob-Gyn informed you that these were potential risks from Pfizer's phase 2/3 trial results? Of course not. No rational individual, especially a pregnant woman, would have considered taking the COVID-19 vaccine in pregnancy had these risks been made known. This rings even more true considering studies by Pineles et al[12] showing that pregnant women had a 75 percent reduced risk of dying from COVID-19 compared to non-pregnant women, and Thorp et al[13] showing no increase in risk of stillbirths with Covid-19 deaths in 2020.

The reality of the situation is that an Ob-Gyn physician became a "trusted messenger" for the government, hospitals, and the medical organizations (ACOG, ABOG, and SMFM), not a "trusted messenger" for their patients—vulnerable pregnant moms and their preborns. The Ob-Gyn physician became nothing more than a bribed government pawn, pushing a deadly experimental mRNA platform while abandoning ethical principles to do no harm. If the Ob-Gyn dared to do their own due diligence and tell their patients the truth, they were threatened with losing licensure, certification, and their employment.

## The Plan to Target the Most Vulnerable Patients—Pregnant Women, Preborns, and Newborns

This is strong evidence that the rollout of the vaccines for pregnant women was preplanned, like the pandemic. Pregnant women were targeted for two obvious reasons: 1) women are the primary health-care decision makers at home, and 2) every single human being on earth could be required to take a "vaccine" if it was shown to be safe, effective, and necessary for pregnant women, the most vulnerable population. Apparently, that was the strategy of those pushing the vaccine. The government/pharmaceutical/military complex with their limitless budget bribed and captured every segment of our society, including the American College of Obstetricians and Gynecologists (ACOG), the American Board of Obstetrics & Gynecology (ABOG), and the Society for Maternal-Fetal Medicine, all of whom continue to push this deadly narrative to this day.

The problem with the mRNA component of the vaccine is that while the spike protein stimulates antibodies it can also trigger a more generalized immune response, resulting in inflammation and deadly over-stimulation of other immune system cells like cytotoxic and myeloid cells. Cytotoxic cells, a.k.a. killer T-cells, are developed in the thymus and destroy cells infected with viruses. Overactive killer T-cells can cause autoimmune diseases, lymphomas, and leukemia. Myeloid cells are a type of blood cell that originates in the bone marrow. They stimulate white blood cells to go to an "infected cell" and destroy it. Myeloid cells are key to maintaining a healthy blood system. Overstimulated myeloid cells can trigger blood clots, strokes, and heart attacks. These are called "immunogenicity" problems—when cells provoke an immune response. Introducing foreign vaccine mRNA into a patient can cause their body to initiate an immune response that triggers a disorder when there was no infection in the first place.

So, as the spike protein delivered by the mRNA attacks the coronavirus, it's also attacking the immune system. A vaccinated pregnant mother produces spike proteins, potentially causing an inflammatory response in every endothelial cell in her body. Endothelial cells line all the blood vessels and regulate exchanges between the bloodstream and surrounding tissues. When spike proteins invade the placenta, they can reduce the nutrients and oxygen the baby gets, causing many potential disasters including stunting of fetal growth, blood clots, fetal death, and many others. Retired neurosurgeon, Dr. Russell Blaylock, warned that "immune stimulation during the third trimester dramatically increases the risk of the child becoming autistic or developing schizophrenia later in life. . . . We will not know if women vaccinated during their third trimester will have children with a higher risk of becoming autistic for at least six years, the usual time span for symptom appearance."[14,15] He also noted that it will take until a child reaches adolescence before schizophrenic symptoms can be observed. Blaylock says women should be warned of these dangers prior to vaccination. Blaylock's description of a world full of autistic and schizophrenic children in the years to come is chilling. But other components of the vaccine are equally monstrous.

To deliver the spike protein into the cells, the mRNA needs transport vehicles that can penetrate the cells. Those vehicles are fatty casings collectively called lipid nanoparticles or LNPs. Think of them as fast cars covered in Vaseline that can slide across virtually every blood barrier in the body. (This was discussed earlier in the book on pages 37 and 131.)

The lipid nanoparticle (LNP) delivery system used for the Covid vaccine was initially designed for delivering cancer medicines and gene therapy. It's used to penetrate the blood-brain barrier and deliver chemotherapy for brain tumors. The blood-brain barrier (BBB) protects the brain from environmental hazards, including medicines and pathogens, like bacteria and viruses. Some estimate that about a trillion lipid nanoparticles are injected into our bodies during vaccination. With each injection, they pass through the blood-brain barrier, the blood-testes barrier, the blood-ovarian barrier, the maternal-placental-fetal barrier, the fetal blood-brain barrier, the fetal blood-testes barrier and the fetal blood-ovarian barrier (as discussed previously). LNPs are specifically designed to break through all the blood barriers and to concentrate in fat-loving tissues like the brain, testes, ovaries, thymus and other crucial organs. A vaccinated pregnant mother passes lipid nanoparticles to her fetus for whom it is an all-out foreign invasion. They land, for example in the thymus gland, the seed of the baby's developing immune system. The thymus literally organizes the fetus's immune system. A poisoned fetal thymus gland results in acquired immunodeficiencies that potentially last a lifetime.

The second layer of the LNP delivery system bringing mRNA to the cells is made up of polyethylene glycol or PEG, also known as ALC0159. It's used in many medications, foodstuffs, and cosmetics. The incidence of severe allergic reactions to PEG is rising as it becomes more common in the environment. When the immune system is overwhelmed with PEG, a severe allergic reaction called anaphylaxis can be triggered that causes the body to go into shock and can be life-threatening. The problem with Covid vaccines is that they deliver PEG into people all over the world with no regard as to how much PEG each person is already carrying in his or her body. The possibility of a person's immune system surpassing its tolerance threshold for PEG and going into anaphylactic shock is there every time a vaccine is administered. Anyone with a known allergy to PEG and/or related emulsifiers should not take the vaccine. Have you seen any warnings about that anywhere?

The gut's reaction to PEG is different from the intradermal or skin reaction. The amount of PEG in breast milk is negligible and below detection, but still present. If the mother has been sensitized and passes on this sensitization in her breast milk to the infant, even if she is not showing signs of sensitization, the infant's immature immune system can trigger a serious reaction after receiving a minute amount of PEG.

Messenger RNA and its lipid nanoparticle delivery system don't just meander into critical organs, they arrive at top speed—at least in rats. In the Pfizer document dump that the DailyClout received, there was a bio-distribution study looking at where mRNA and liquid nanoparticles went in lab rats after being injected with 50 micrograms of Covid 19 vaccine. A shockingly rapid accumulation of liquid nanoparticles was seen in the rat's ovaries. In the first fifteen minutes, the amount of LNP/mRNA registered in the rat's ovaries was 0.104 micrograms. After forty-eight hours, it was 12.261, or almost 118 times higher. The study was stopped after forty-eight hours, raising the deadly spectre of an exponential rate of accumulation beyond the forty-eight hours. High concentrations of lipid nanoparticles were also found in the rats' livers, spleens, and adrenals.

A regular Covid vaccine is about 30 micrograms, or 20 micrograms less than the rats received. Does getting a lower dose of lipid nanoparticles matter? The *Pfizer Documents Analysis Reports* points out that "Any evaluation of the safety of this delivery system for a vaccine needs to evaluate whether penetration of the blood barrier by the lipid nanoparticle delivery system conveys its own harm. Studies have proven that ENMs (engineered nanomaterials) that can cross or bypass the blood-brain barrier and then access the central nervous system, carry the potential of neurotoxicity. This evaluation was never done in the Pfizer safety and efficacy trials. Therefore, it is impossible to know whether the vaccine is safe in this arena. Pfizer did not prove the safety of the nano-lipid delivery system for the brain." In addition, "When the government and scientists told us over and over again at the outset of the vaccination program that the injections would stay in the deltoid shoulder muscle, they were lying and knew it. Ever since lipid nanoparticle delivery systems were patented more than two decades ago, it was known that they enter the body's circulation system and can find their way to many end points. It's also known that spike proteins, mRNA and nanoparticles stay for weeks to months and possibly years in human tissues and the harms from these agents are being identified almost daily."[16]

No proper safety testing of the vaccine's delivery system has been done even though deaths and extreme adverse events happening everywhere and in people of all ages indicate there are serious problems—not just "signals."

## The Case of Cody

Cody Hudson's experience with lipid nanoparticles was more than just a "signal" to him. A vibrant, athletic young man just entering his twenties, he took two doses of the Pfizer vaccine in 2021. Children's Health Defense reporter, Brenda Balletti, wrote a searing piece describing Cody's vaccine nightmare: "Healthy 21-year-Old Given 3 Days to Live After Pfizer Shots Led to Rare Autoimmune Disorder."

Cody had flu-like symptoms after the first dose and got a rash on his arms after the second. He thought the rash was eczema. Then he began to have aches and pains, including severe knee pain. While the pain and aches worsened, he began coughing up blood. "At three in the morning," his mother, Heather, recalled, "he was coughing up blood and hanging onto the walls trying to stand up and his face was swelling." "I couldn't lie down because it was so painful," Cody added. "I felt like I was going to die from the pain. I don't know how to describe it."[17] They went to the emergency room where he was diagnosed with a pulmonary embolism in his left lung, numerous small blood clots in his right lung, and thrombocytopenia (a low platelet count that causes bleeding). Cody was bleeding and clotting at the same time. That's when they gave him three days to live. Heather Hudson, Cody's mother, did an extensive amount of medical research and she initiated the autoimmune blood testing that later confirmed his diagnosis of antiphospholipid syndrome (APS).[18] After five harrowing months of trying to keep Cody alive, his mother took him to see internal medicine physician, Dr. Eduardo Balbona. He then confirmed the diagnosis of APS, an autoimmune disorder that occurs when the immune system mistakenly attacks normal proteins in the blood, causing increased blood clotting. Balbona explained to the Hudsons that Cody's autoimmune reaction was due to lipids in the mRNA vaccine.

Cody still suffers from autoimmune issues that he'll most likely have for the rest of his life. Lesions appear and disappear on his hands and arms. He has random nerve pain and myopathy—damaged muscles that restrict his movement and cause weakness. He also has neuropathy, nerve damage that causes numbness and tingling in his feet and hands, burning, stabbing or shooting pains in affected areas, loss of balance and coordination, and muscle weakness.

At twenty-two, Cody walks with a cane and probably will for the rest of his life.

In a peer-reviewed case report he wrote on Cody, Balbona warned that he wasn't the first to see this problem: "Others have noted that 'antiphospholipid antibodies' may represent a risk factor for thrombotic [clotting in veins or arteries] events following COVID-19 vaccination and deserve further investigations." "They also propose that in the case of 'pre-existent aPLs' [antiphospholipid or clotting disorders], infections may trigger pro-inflammatory cascades able to promote development of a full-blown antiphospholipid syndrome."[19]

Balbona also mentioned other serious autoimmune reactions occurring after vaccination, including New-Onset Systemic Lupus Erythematosus (SLE), an autoimmune disease causing the immune system to attack its own tissue and damage the affected organs, and acquired hemophilia (when the blood can't clot properly). He also described how a forty-two-year-old woman with no previous medical history, took the vaccine and got both lupus and antiphospholipid syndrome. She suffered from the same joint pains Cody did, had difficulty breathing, and suffered from hypoxia—when the body is starved for oxygen. "These issues are similar to the multi-system symptoms that occurred in the case [Cody] presented herein," Balbona noted.

If the vaccine's lipid nanoparticles and other components could bring down a strapping twenty-two-year-old man and a healthy forty-two-year-old woman, it's no leap of logic to assume that an increased level of death and devastation was in the cards for the mothers and babies in my practice.[20] That was the catalyst for my writing this book. The statistics in my office skyrocketed right along with those around the world. In the US, the miscarriage rate jumped to 81 percent, stillbirths went up 5.8-fold, neonatal deaths, up 7.9-fold, and there was a 14.7 percent rate of complications in breastfed newborns. Among US military personnel, the Armed Forces Health Surveillance Division's Medical Epidemiology Database recorded a mind-boggling 74 percent rise in congenital malformations in less than a year after Covid vaccinations began in 2021.

Dr. Naomi Wolf, of DailyClout, reported that "In vaccinated countries, babies are dying in unprecedented levels. In Ontario, Canada, a doctor there said that usually they have five or six neonatal deaths, and in a three-month period that they had 86. This was his testimony. In Scotland, which is an almost completely vaccinated country, there's double the number of babies dying. . . . And in Israel, also a highly vaccinated country, Rambam hospital has 34 percent [more] spontaneous abortions, miscarriages and newborn deaths to vaccinated mothers than unvaccinated."[21]

What I saw in my practice beginning in 2021 matched the previously mentioned adverse effects triggered by Covid vaccines. I saw increases in spontaneous premature deliveries and medically necessary preterm births. I saw severe placental abnormalities, specifically calcifications or premature aging of the placenta, lacunae that cause high-velocity blood flow leading to the placenta growing outside the uterus and invading surrounding organs, and infarcts, yellowish white proteins that clog placental arteries and impair fetal blood circulation. I also saw placental bleeding, placental abruption—when the placenta separates from the uterus—fetal growth restriction, and severe early onset preeclampsia, a dangerously high blood pressure condition.

In addition, there was a substantial increase in malformations of all organ systems of embryos and fetuses. I saw congenital cystic hygroma or abnormal growths on babies' heads and necks; hydrocephalus, an accumulation of cerebral spinal fluid in the brain; neural tube defects—the neural tube forms the early brain and spine; cardiac defects and congenital diaphragmatic hernias. Many of these occurred in mothers who had been vaccinated during their first trimester, the most delicate time of pregnancy. Spontaneous abortion rates are also much higher during this phase. So are the numbers of premature babies, babies who are small for their gestational age, and babies with congenital abnormalities. Neonatal deaths are also higher.

There were also severe post-partum hemorrhages. In some cases, post-partum hysterectomies had to be performed, a devastating outcome for mothers who had hoped for more children. I saw immune complications in the vaccinated mothers as well as a substantial increase in risk of immune and autoimmune complications in their newborns/infants.

I am certain that these outcomes are the results of the use of the most highly inflammatory substance ever injected into humans, particularly given the unprecedented increase I saw in babies born with swollen lymph nodes, a sign of lymphadenopathy, auto-immune disease, or cancer. The placental calcifications are also telling. They occur in human tissues experiencing an inflammatory process leading to necrosis or cell death.

There is no other way to sum this all up but to say that we were in a killing field.

I am appalled that those babies of vaccinated mothers who died in the womb, at birth, or right after birth have not been systematically autopsied and studies done to determine if and how the vaccine may have contributed to their deaths. A study by Ob-Gyn physician Jian-Pei Huang titled "Nanoparticles Can Cross Mouse Placenta and Induce Trophoblast Apoptosis,"[22] suggests how the vaccine might kill a fetus. Dr. Huang found that nanoparticles crossed the placenta, entered the fetus, and caused the death of trophoblast cells that normally provide nutrients to the developing embryo and then become a big part of the placenta.

Wolf says looking into lipid nanoparticles and their association with these deaths should be an immediate priority:

> I want to go to the amniotic sac, the placenta. This is a baby's home for nine months and it's super delicate. These lipid tiny hard fatty casings are entering, traversing the placenta. And personally, I don't yet understand how something can traverse a membrane without harming or weakening or altering a membrane. They haven't explained that to me yet. But even say the placenta is intact with these little particles traversing it, they are in the fetal environment, they're in the amniotic fluid that the baby depends upon to create a healthy environment for it to grow. And they're traversing; they're in every cell of the body according to the manufacturer's earlier website. They've deleted this, but that means . . . they're in the baby itself. . . . Understanding how lipid nanoparticles work, the fact that babies are dying is an incredibly urgent

emergency for us to stop everything and immediately figure out why they're dying.[23]

A baby born who survives in a vaccinated mother's womb still isn't out of the woods after birth, if the mother is breastfeeding. A five-month-old infant breastfed by a mother who had just received a second dose of the Pfizer vaccine died of clotting. The day after the mother took the shot, her doctor reported to VAERS that "her breastfed infant developed a rash and within 24 hours was inconsolable, refusing to eat and developed a fever."[24] In the emergency room, tests showed the infant had elevated liver enzymes; a sign of liver damage or inflammation. The baby died two days later after being diagnosed with thrombotic thrombocytopenic purpura, a blood disorder where clots form in small blood vessels throughout the body, blocking blood flow to the organs. The blood clot theme—a hallmark of adverse events connected to lipid nanoparticles—repeats itself over and over in adverse Covid-19 vaccine events.

But it doesn't end there with the lipid nanoparticles.

The lipid nanoparticle trail of death and destruction even reaches into the future. In preborn babies and adult women, lipid nanoparticles have caused catastrophic reductions in the number of eggs in the ovaries, of which there are only a limited number of about one million at birth. After birth, there is a natural die-off of the eggs that is irreversible and continues throughout a female's reproductive life until they are depleted. This process is called apoptosis. LNP toxic substances kill the eggs and accelerate this natural die-off. The consequences on future fertility from this are dire, whether in a female fetus or a woman of reproductive age. Worldwide, the premature depletion of human female egg reserves is happening. Call me paranoid, but I can't help but look at this as a deeply pernicious aspect of clandestine efforts to drastically reduce the world population.

The depletion of female egg reserves, which interferes with conception, is only part of the vaccine's "360-degree attack on human reproductive capability." I've seen a dramatic rise in couples seeking advice for infertility since the 2021 vaccine rollout. Both men and women have issues. Men have problems with loss of libido, erectile dysfunction, abnormal sperm motility, and low sperm count. This is not unexpected as the vaccine's lipid nanoparticles are known to concentrate in the ovaries and testes. In an autopsy documenting a Covid vaccine–caused death, Dr. Arne Burkhardt documented complete loss of sperm production and testicular destruction by spike protein in a twenty-eight-year-old male who had previously fathered a child.

The CDC/FDA's Vaccine Adverse Events Reporting System or VAERS database documented cases of "infertility," "female infertility," "male infertility," and "abnormal fertility tests" connected to Covid shots over a period of 2.3 years, to flu vaccines over 33 years, and to all other vaccines except COVID-19 shots over 33 years. The results were astonishing. There were 171 cases of infertility from the COVID-19 shots in 2.3 years compared to one case of infertility in 33 years for the flu vaccine and 121 cases for all other vaccines over 33 years.

"I can definitely say since the vaccine rollout started we have seen in our practice a decrease in new OB [pregnancy] numbers, which would be a rise in infertility by 50 percent," observed Ob-Gyn doctor Kimberly Biss. "We've also seen an increase in miscarriage rate of 50 percent, and I would say there's probably a 25 percent increase in abnormal pap smears as well as cervical malignancies in our area." Dr. Biss practices in St. Petersburg, Florida.[25]

Menstrual abnormalities are also through the roof since the vaccine rollout but so far, the government and the vaccine manufacturers are downplaying or ignoring it altogether. Research analyst and founder of the MyCycleStory Group, Tiffany Parotto, took it upon herself to create a platform dedicated to "supporting independent research, community and solutions for the fertility and menstrual irregularities we are being faced with." In spring 2021, says Parotto, "women began sharing their experiences of significant menstrual cycle changes on social media, only to have their conversations suppressed and deleted. In response, a team of researchers, doctors and experts conducted the MyCycleStory survey to investigate these irregularities and amplify the voices of affected women."[26]

Before MyCycleStory did their survey, a previous survey of women launched in April 2021 had more than 150,000 respondents. MyCycleStory followed up on that survey and gathered a wider array of general and menstruation-related symptom data.

The survey showed that more than 90 percent of the respondents did not experience menstrual abnormalities until after the 2021 vaccine rollout. Most notably, 292 women (4.83 percent of the sample) reported having experienced symptoms consistent with a severe form of abnormal vaginal bleeding called decidual cast shedding. This rare occurrence is described on MyCycleStory's website: "Decidual cast shedding can happen to some women during their menstrual cycle. Normally, the lining of the uterus, called the decidua, gets thicker to prepare for a possible pregnancy. If pregnancy doesn't occur, the decidua sheds and comes out of the body through the vagina. In the case of decidual cast shedding, the decidua comes out in one solid piece that looks like the shape of the inside of the uterus, almost like a triangular mold. This can be a rare and sometimes painful experience for women. It's important to remember that that everyone's body is different, and not all women will experience decidual cast shedding."[27] To put pre- and post-vaccination instances of decidual cast shedding into perspective, the survey identified 292 cases in 7.5 months in 2021, compared to fewer than 40 cases published in the medical literature over 109 years.

The stories of excessive bleeding and odd menstrual behavior on the MyCycleStory website are horrendous, but those that end with the termination of a woman's reproductive capabilities are particularly disturbing.

Dr. Naomi Wolf has been banging the drums on this issue, sounding alarm after alarm, yet all remains quiet at the FDA. The NIH has given the issue a nonchalant nod by acknowledging that vaccinated women's menstrual cycles have extended on average an extra day a month. But it's nothing, they say, just tiny blood vessels being impaired by the contents of the vaccines. Wolf vehemently disagrees:

> That's twelve extra days of bleeding on average for a year. Almost half a month of extra bleeding. And, of course, extra bleeding means, you know, that uterine lining is there for a reason and it sheds for a reason, and these things are very delicate hormonally. It's not like you can just have an extra day of bleeding on average a month and nothing else is wrong. That's a huge red flag . . .[28]

Another huge red flag is the fact that vaccinated women are affecting the menstrual cycles of unvaccinated women. I was a coauthor on another MyCycleStory survey of 6,049 women that focused on 3,390 unvaccinated women who had not taken COVID-19 vaccine but had been in close contact with someone who had been vaccinated. Within three days, 50.1 percent of the women had irregular periods. Within a week, 71.7 percent.[29] The study showed that daily proximity to vaccinated individuals increased the risk of heavier bleeding and changes in menstrual timing, suggesting a potential link between vaccine substances and menstrual changes in unvaccinated women.

The exact substance being shed from vaccinated to unvaccinated women has not been proven and could include a variety of candidates, including the spike protein and lipid nanoparticles. Pfizer clinical trial participants who were given the vaccine were advised to report any exposure they had to a pregnant woman by way of inhalation or skin contact or to the woman's sexual partner prior to the time of conception. This begs an urgent question:

Why didn't Pfizer advise the public about this and issue a warning about this kind of contact? Clearly Pfizer knew about the shedding phenomenon when they instructed those recently vaccinated in their clinical trials to NOT have intimate sexual relations with pregnant women. It's an unspeakable omission.

Both men and women's reproductive systems are under attack, but women are far more affected. Pfizer documents show that women experience about three times more adverse events than males. The specific risk to women's reproductive organs and their functions is even greater. Pfizer's reissued "5.3.6 Cumulative Analysis of Post Authorization Adverse Event Reports ... Received Through 28-Feb-2021,"[30] shows the categories of adverse events and breaks down the percentages between males and females. The contrast is stark.

In every category, females substantially outnumber males. Examples: Autoimmune: women 81%, men 19%; Cardiac: women, 77%, men 21%, Dermatologic: women 94%, men 6%; Musculoskeletal: women 80%, men 20%. But the whopping difference between the sexes is in the reproductive system adverse events category: women, 148,874, men: 1,745.

Let's sit for a minute with the reality of shedding. We now live in a world where someone can decide to put a dangerous substance in their body and just by being near that person, our bodies can take in that substance and be affected by it in the most profound way possible—by damaging our reproductive capabilities—without even knowing it. The notion of a DNA-altering vapor excreting from the skin of the people we are around, going into our bodies whether we want it or not, is like a terrifying, sci-fi novel. And yet it's real and has been proven beyond any doubt. Could there be anything more bleakly totalitarian?

How about having pieces of the vaccine embedded in our DNA to pass on to all future generations? What will that DNA tell the cells of future generations to do? Nothing good, you might imagine, and you'd be right.

Microbiologist Kevin McKernan was the first to discover DNA contamination in both Pfizer and Moderna Covid vaccines as well as a cancer-causing agent known as SV40 in Pfizer's vaccine.[31] This prompted the World Council for Health to organize an Urgent Expert Healing conference on October 9, 2023. The next day, they put out a press release headlined: "World Council for Health Expert Panel Finds Cancer Promoting DNA Contamination in Covid-19 Vaccines: International expert panel concludes that Covid vaccines are contaminated with foreign DNA and that SV40, a cancer-promoting genetic sequence has been found in the vaccines."[32]

Their Summary of Findings absolutely stunned me, which is why I'm reproducing it in its entirety here:

1. Bacterial DNA (plasmids) has been found in mRNA vaccine vials.
2. A cancer-promoting genetic sequence—SV40—has been found in the Covid-19 vaccines. This was not present in the vials used for the approval studies but has been found in all vials of the BioNTech vials disseminated for public use.
3. These discoveries have been confirmed in several independent laboratories worldwide.
4. The discovery was originally made in April 2023 by Kevin McKernan, at which point regulatory bodies were contacted. No official reply has been received.
5. Multiple mechanisms exist in which this genetic information might be integrated into the human genome.
6. This DNA could instruct our bodies to produce mRNA and foreign proteins for an unknown period with potential implications for subsequent generations.
7. There is no constructive purpose identified for the undeclared SV40 promotor sequence, which in addition to its cancer risk, enhances the capacity to incorporate the other foreign genetic material into the recipients' own chromosomes potentially rendering them (and possibly even their offspring) permanently genetically modified.
8. There are multiple completely undeclared genetic sequences in both Moderna and Pfizer vials with the SV40 sequence found only in the Pfizer vials. However, latent SV40 infections in a significant portion of the population could present the same SV40 risk to Moderna recipients.
9. Even in the absence of chromosomal integration, the DNA plasmids could generate mRNA for the spike protein toxin and other harmful proteins for prolonged and unpredictable periods of time.
10. Integration of foreign DNA into the human genome disrupts existing natural genetic sequences; this carries further risk of disease including cancer.
11. The Covid-19 vaccines qualify as GMO (genetically modified organism) products, which require approval in addition to that required for older, more traditional vaccines.
12. Informed consent for these products is impossible as the risks of the products have never been formally and transparently assessed by

regulators and are not fully known. Independent assessment of the emerging and available evidence is that these products are extremely dangerous with implications for disease, death, transmission, and inheritance.
13. An immediate moratorium on these novel genetic "vaccines" was demanded by the expert panelists.

In response, Maggie and I published an article entitled "A Call for Immediate Moratorium on the use of COVID-19 Vaccines in pregnant women" on March 3, 2024.[33] We focused on a recent study by Lin[34] published in the *American Journal of Obstetrics & Gynecology (AJOG)* 2024, entitled "Transplacental transmission of the COVID-19 vaccine messenger RNA: evidence from placental, maternal and cord blood analyses postvaccination," found that COVID-19 vaccine mRNA was detected in the placentas of two pregnant mothers who had been vaccinated with Pfizer's mRNA COVID-19 vaccine shortly before delivery. The study also found that spike protein expression was detected in the placental tissue of the earlier-in-time vaccinated mother, demonstrating bioactivity of the COVID-19 vaccine mRNA after reaching the placenta.

Also of great concern, vaccine mRNA was detected in the fetal blood of the only patient sampled, documenting a 100 percent rate of transmission in their study. The implications of these research findings are profound—they indicate that COVID-19 mRNA vaccines penetrate the fetal-placental barrier and reach the intrauterine environment, where the mRNA can then be translated into spike protein and expressed in the placental tissue. In the results, Lin and colleagues stated the vaccine mRNA had a "notably high signal in the decidua." Then if the vaccine mRNA is concentrated in the decidua—the portion of the lining of the uterus that is in closest proximity to the preborn—it likely produces high concentrations of spike protein, thus explaining the heavy abnormal menstrual bleeding and the severe complications in pregnancy. The spike protein is an extremely inflammatory substance and definitely contributes, at least in part, to the severe complications in pregnancy caused by the COVID-19 vaccines.

Suddenly, all these reports of turbo cancers, particularly in young people, take on a new light. Japanese Professor Murakami of Tokyo University describes the discovery of SV40 as a "staggering problem" for Pfizer. "The question is why such a sequence that is derived from a cancer virus is present in Pfizer's vaccine," he said. "There should be absolutely no need for such a carcinogenic virus sequence in the vaccine. The sequence is totally

# Vaccine Masquerade

unnecessary for producing an mRNA vaccine . . . That is not the only problem. If a sequence like this is present in the DNA, the DNA is easily migrated to the nucleus. So, it means the DNA can easily enter the genome."[35]

"This is such an alarming problem," Murakami added, "It is essential to remove the sequence. However, Pfizer produced the vaccine without removing the sequence. This is outrageously malicious."

The fact that SV40 was not in the vaccine vials submitted for approval studies but was in the vials used for injecting the public can't be an oversight or a coincidence, particularly since it's not necessary for producing the vaccine. So, I'll call it what it is—a crime against humanity.

That crime is compounded by the fact that early on, effective early treatment protocols for Covid were available before the vaccines were rolled out. A massive, fraudulent campaign against the safe drugs that were part of those protocols, specifically ivermectin and hydroxychloroquine, and the vilification of the physicians who prescribed them to their Covid patients, resulted in a much higher death rate globally. What has since come into sharp focus is the fact that Dr. Anthony Fauci, who was in charge of the US government's Covid response, had other more pressing priorities than saving Americans from the virus with early treatment protocols. Many Americans died unnecessarily as he and his accomplices waited for the gravy train of their lives—the vaccines, and the massive profits they would provide.

FDA heroine, Dr. Frances Kelsey, must be rolling in her grave right now. She saved the United States from the horrors of thalidomide by insisting that the drug's manufacturers prove their drug was safe and effective. They couldn't, so she didn't approve it. Where are the FDA's Frances Kelseys now? They've disappeared deep down the pockets of the pharmaceutical industry, leaving all of us at the mercy of these intergenerational killers. What the FDA accepted from the Covid-19 vaccine manufacturers as proof that their products were safe and effective were fig leaves—a few studies that didn't even bother to hide their fraudulence, their sins of omission and commission.

Starkly put, in the case of Covid vaccines, the FDA has literally taken our tax dollars while allowing the pharmaceutical companies to maim and kill us.

And, to paraphrase Dr. Wolf:

*It's eerily creepy how focused they are on reproduction.*

## A Good Birth: America 1953

*The innocent and the beautiful have no enemy but time.*
—W. B. Yeats, Irish poet and playwright

Picture an unassuming, low-budget apartment in Detroit, Michigan, on a quiet, tree-lined street. Arriving home from work on April 16, 1953, my father, Ken, had been expecting to drive my mother to the hospital to give birth to their second child—me. He came through the kitchen door, paused, and wondered why a steak was thawing in the sink. Being of modest means, my parents rarely had steak for dinner.

"Mollie?" he called out.

"Ken come in here, I'm in the bedroom!" she called back.

My father entered the disheveled bedroom, only to unexpectedly find my mother sitting up in bed nursing a baby. Me.

Feeling rather stunned, he was not able to put the pieces together.

"Mollie, what in the world is going on here? You're already back from the hospital? Why didn't you call me? And why are you thawing a steak for dinner? What is going on, are you OK? This makes no sense!" My poor father was so upset and confused he needed the whole series of events clarified before he could focus on his newborn son.

My mother confessed that I'd been born that afternoon, with the help of one of her nurse friends. She had co-conspired with her friend to deliver the baby at home, to avoid a second hospital birth, as she had terrible memories of the first one. When my older brother was born, they had placed her under twilight anesthesia, and she had no recollection whatsoever of his birth.

"That's not steak Ken, it's a placenta," she said, trying to conceal her smile. "Now come over here and meet your son. It's a boy, what you were hoping for!" Finally, after his initial shock, my father came over and got a look at me, remarking how strong and healthy I looked.

Grateful for both my parents, I'm acutely aware of how much has changed—how our *innocence* has been lost—how I could never have conceived of the Covid nightmare that would descend on all of us, decades later.

I also did not know, at the time, that I would follow in my mother's footsteps. A strong, devout Irish Catholic woman, she worked as a labor and delivery nurse in Detroit, where I grew up. My mother had a passion for pregnancy and childbirth, as evidenced by the bookshelves that were filled with fascinating reads on those subjects. These books had a profound influence on me while growing up.

# Vaccine Masquerade

My father, Ken, was a mechanical engineer, focused on metallurgy. He was a hard-working man, and tough World War II veteran, with a great sense of humor, work ethic, and strong family values.

We were not well off, but my mother created a comfortable and inviting home. As soon as I learned to read, I began exploring her bookshelves—homing in on her vast collection of medical and historical obstetrical books. She was particularly fascinated by the works of the legendary and heroic Hungarian physician, Ignaz Semmelweis, and I inherited her love for him. She would tell me stories about how he solved the mystery of what was killing so many women on a delivery ward in 1860s Vienna.

My mother told me of how Semmelweis persisted in solving the puzzle of those deaths—how he figured out the true cause of the deadly wave of puerperal fever that was killing those birthing mothers. It turns out that the answer was fairly simple: Semmelweis identified that "cadaverous particles" were being transmitted by doctors' dirty hands, between performing autopsies and child deliveries. To confirm his thesis, he promptly went to work to develop a cleansing solution that, when used, immediately solved the problem of the mothers dying in childbirth.

Semmelweis's problems, on the other hand, were just beginning.

The hospital wards in question had been set up to address the issue of infanticide of illegitimate children. Many of the women who gave birth in those wards were poor; some were prostitutes. In exchange for free medical care, they would be test-subjects for doctors and midwives in training.

There were two maternity clinics at the Vienna General Hospital—each with about 3,500 deliveries per year. The first clinic had a death rate from puerperal fever of about 20 percent while the second averaged about 1.7 percent That's an enormous difference. At the clinic with the higher death rate, doctors were being trained, and their training included autopsies. At the second clinic, it was all women, training only in midwifery.

Birthing mothers begged, sometimes on their knees, not to have to deliver their babies on the "death ward," which they'd heard so many horror stories about. Some even chose to give birth on the street rather than risk going to the frightening clinic. Semmelweis documented how this pitiful sight devastated him.

He began to puzzle over why even a street birth was safer than a birth at the cursed clinic. He eliminated theories by trial and error—some were quite strange, including an early theory that the women were going into shock by the loud sound of priest's bells in the corridor. He asked the priests to take a different route, but it made no difference.

One by one, Semmelweis eliminated various theories. A popular theory at the time was a fiction known as "miasma" which insisted there was some kind of mysterious substance in the air that made the mothers get sick and die. Another theory was "climate." (Some things never change.)

Around this time, Semmelweis wrote, "When I look back upon the past, I can only dispel the sadness which falls upon me by gazing into that happy future when the infection will be banished. . . . The conviction that such a time must inevitably sooner or later arrive will cheer my dying hour."[36]

Semmelweis's big breakthrough came in 1847, following a personal tragedy. His good friend and colleague, Jakob Kolletschka, had been accidentally cut with a student's scalpel while performing a post-mortem on a woman who had died of puerperal fever. Kolletschka immediately fell ill, with the identical symptoms, and soon thereafter he died. Semmelweis proposed that "cadaverous particles" from the scalpel Kolletschka had been using, were the cause of his death.

Determined to confirm this theory, Semmelweis created a solution of chlorinated lime that doctors, nurses, and students began using—to wash and sterilize their hands after autopsies, instead of going straight from post-mortems to childbirth wards without any decontamination. When they did this, the deaths from puerperal fever stopped.

But ironically, that was when Semmelweis's real problems began. What happened?

The aristocratic gentlemen class of physicians in Vienna at that time did not take kindly either to the idea that their hands were dirty, nor that they had killed all those women, albeit unintentionally. They began, little by little, to gang up on Semmelweis, attacking him for various things, including not presenting his findings clearly enough in the scientific literature for them to be able to judge whether it was sound and true. Semmelweis, in response, grew ever more combative and belligerent, excoriating his colleagues for murder.

According to Wikipedia:

> "The maternal mortality rate dropped from 18% to less than 2%, and he published a book of his findings, *Etiology, Concept and Prophylaxis of Childbed Fever*, in 1861.
>
> "Despite his research, Semmelweis's observations conflicted with the established scientific and medical opinions of the time and his ideas were rejected by the medical community. He could offer no theoretical explanation for his findings of reduced mortality due to hand-washing, and some doctors were

offended at the suggestion that they should wash their hands and mocked him for it. In 1865, the increasingly outspoken Semmelweis allegedly suffered a nervous breakdown and was committed to an asylum by his colleagues. In the asylum, he was beaten by the guards. He died 14 days later from a gangrenous wound on his right hand that may have been caused by the beating.

"His findings earned widespread acceptance only years after his death, when Louis Pasteur confirmed the germ theory, giving Semmelweis' observations a theoretical explanation, and Joseph Lister, acting on Pasteur's research, practiced and operated using hygienic methods, with great success."[37]

After a protracted series of battles, in which his colleagues dug in their heels—all while Semmelweis grew more combative—they began to deride and mock him even though they knew he was right. What he was saying was so simple. He summed it up in three words at the opening of a presentation once gave: "Wash your hands."

But they did not want to listen, and eventually they even stopped washing their hands with the chlorinated lime solution. Unable to cope with the stress, ridicule, and eventual isolation from his colleagues, Semmelweis's drinking escalated, and he is said to have begun to visit prostitutes.

Then, by way of trickery, he was committed to a mental institution by his colleagues. There, he was beaten shortly after arrival, and within two weeks, he had died of sepsis.

Semmelweis is sometimes credited with being the inventor of "germ theory," and even today, in a blatant display of hypocrisy, the globalist leaders at the World Health Organization (WHO) can be seen singing his praises and trying to appropriate his legacy.

What I admire most about Semmelweis was his determination to help his patients, who were among society's least advantaged, to say the least. But also, I admire his refusal to go along with fashionable wrongness in medicine.

Semmelweis set out alone, ignoring the dominant theories and cliques, and found out why birthing mothers were dying in such numbers on that particular ward. Nobody else seemed to even care. He then proved it, by eliminating the cause, and stopping the deaths. Yes, he pioneered antiseptic medicine, but he also pioneered fighting superstition and nonsense in science. He certainly could be called the pioneer of many things, but one that particularly stands out in my mind, would be common sense in medicine.

Semmelweis clearly believed in taking full responsibility for determining why his patients were dying. He didn't dismiss it or turn his back. It haunted him, and he confronted the issue, and eventually, solved it. The eponymous

"Semmelweis Reflex" refers to the automatic, robotic tendency to reject new evidence or knowledge because it contradicts existing, established beliefs or norms. It describes the herd-like mentality of so many doctors and scientists who cling for decades to bankrupt ideas, rather than considering something new. This herd-like mentality causes countless unnecessary deaths.

Sometimes in my hours of darkness, in these totally insane Covid years that have turned medicine upside down and inside out, I imagine a conversation with Semmelweis. I imagine that I can hear him sharing his unshakable values and guidance, and I find that comforting.

I also have a good sense now what it means and what it feels like to go through all the training needed to become a doctor, only to find yourself shunned. Shunned, fired, alone.

I also know, with 100 percent certainty, that I'm right.

In my certainty, I have offered to debate any of them, any time, under any conditions. But, I've learned, quite painfully, how this game is played. They just ignore me—place me in a dome of silence, as though we had never known each other or had ever worked together.

# Chapter Thirteen

# The Turning Tide: A Story of Hope and Perseverance

*Whoever welcomes one of these little children in my name welcomes me; and whoever welcomes me does not welcome me but the one who sent me.*

—Mark 9:36–37

*If we want this thing to end, companies will be responsible for the products that they make, period. And then they'll stop. The minute Pfizer, and the minute Moderna have fiscal and civil and criminal liability for the things that they actually are producing, we will not have a pandemic.*

—Dr. David Martin, *InfoWars* interview, June 21, 2024

It's been four and a half years since the beginning of my Covid nightmare—a nightmare that befell each and every one of us, worldwide.

However, within that nightmare, I've met, and continue to meet, countless remarkable and courageous people. Many have lost loved ones to the shots; many of those have been my patients. Many have inspired me to continue speaking out, despite the risks. I've gained new, trusted friends, and lost quite a few. I've been taken to my knees, been fired, censored, labelled, and sued; but I remain steadfast in telling the truth about these shots, and if I can wake up just one soul, it has been worth it.

One young mother, who I met while filming a documentary, tells her story below. She is a voice for the many mothers who have lost their babies—some in-utero, some post-birth. But Victoria refused to give in to the death and despair that surrounded her. Rather, she used her light—the light from her Creator—to stand up and fight for this blessed gift we call Life.

## An Angel Who Lived for Eleven Hours

Victoria White, a twenty-seven-year-old mother of then two-year-old daughter, was employed at a nursing home in her hometown of Peterson, West Virginia, when vaccine fever and propaganda finally caught up with her. She had stalled as long as she could, but to keep her job, she realized she had to get the Covid injection.

Just before she went to get the first shot, Victoria (Tori) discovered she was pregnant with her second child. She was assured, by the clinic where she got the shot, that it was "perfectly safe," and, trusting that, she received two injections, two weeks apart, both in the first trimester of pregnancy.

Tori could feel it right away—she had strange pains and other unusual symptoms throughout the pregnancy—it was not at all like her first one, which had gone very smoothly, no issues.

When it came time for Tori to give birth, she was induced, at term.

Relaying her story in the 2023 documentary, *Shot Dead* [directed by Teryn Gregson], Victoria, or Tori, as she is known, describes an ominous feeling she had when she was in labor. The first alarming sign was that her baby's heart rate plummeted and stayed way down for eight whole minutes. Immediately after her daughter, who she named Naomi, was born, a nightmare unfolded. The baby was subdued, dejected, and purple from lack of oxygen.

"She maybe cried one time," Tori said, sitting in a leather sofa with her father, mother, and first child, a young daughter. "She laid there, and I looked at her. She was looking back at me and . . . she kept turning purple and just wasn't getting enough oxygen. They let me give her a kiss and then the nurses took her away and said, 'She'll be fine, she just needs oxygen, maybe some IV fluids'."

Tori's voice breaks as she struggles to speak, now openly weeping at the memory. "I kind of knew when they put her in my arms, there was no chance. Her body was so heavy, and her mouth was purple and . . ." She shakes her head and wipes away tears. "It wasn't what I wanted for her."

When she says these simple words, it hits you, the tragedy: For a mother to see her newborn baby's suffering, to see it in her eyes, and to say it so simply and so poignantly: "It wasn't what I wanted for her."

# The Turning Tide

Tori was given only a very brief moment with her newborn daughter. The nurses removed Naomi and took her to the NICU. "That's the last time I saw her," says Tori.

Naomi lived for eleven hours.

What went wrong? Her umbilical cord was very short, perhaps 10 cm, instead of 30. The placenta was ". . . the size of a baseball," Tori's mother, Rhonda, said. "She had pustules on her head that were leaking blood and fluid. Her body was acidic. One lung had collapsed completely, and she was diagnosed with a diaphragmatic hernia."

It is beyond tragic, to witness the suffering of a newborn in this condition.

"I've spoken a lot with Tori . . . we've spoken about the baby that she lost," Dr. Thorp says in the film. "It's fresh. She told me that her baby was diagnosed with congenital diaphragmatic malformation. That's a very serious malformation. If that malformation is not identified prior to birth, we're likely going to lose that baby."

Stunningly, though this is a vanishingly rare condition under normal circumstances, there was another baby delivered with the same condition, at the same hospital where Tori gave birth, that same night. "Earlier that night there was another girl who had the same issue, but her baby ended up surviving," says Tori.

Tori's mom, Rhonda, adds that the doctor told the family that they usually see one of these in 2,500. "Before this, he had seen one in his whole career of thirty or forty years," she says, "but he had seen three in the last month." A frightening reality.

Naomi is buried in the family grave plot, next to her great grandparents. Her tombstone reads:

*Naomi Lainey White*
*1/15/23 - 1/16/23*
*Please Step Softly, Our Baby Sleeps Here*

"She's our little angel. She was beautiful," says grandfather, Jeff, who visits her grave most days.

Despite the horror and heartbreak, there was, for Tori, a bright light that emerged from the tunnel.

Tori got pregnant again, in 2023, and gave birth to a boy named Ares, in the late winter of 2024—Ares Ryker James White. I had stayed in close contact with Tori over the course of this pregnancy and feel so honored to have been the inspiration for her son's middle name.

On a Zoom call, Celia and I chatted with the extended family as they sat on their porch on a spring day in 2024—Tori holding the sleeping, swaddled baby in her arms, and Jeff, her father, next to her. "I didn't have my hopes up," says Tori quietly. "I tried to . . . not be as excited as I had been with my daughter, which is actually kind of sad."

"Hearing Ares cry for the first time brought tears to my eyes," says Jeff. "It was an amazing feeling after what we've been through."

This time, the family had chosen a different hospital system. No one had mentioned Covid shots to Tori, though they did try to push RSV, briefly, but she was now alert to the sound of the "script"—the "safe and effective" script.

## People Waking Up, One Soul at a Time

As someone who has been in the thick of this Covid storm since the beginning, I wondered out loud if their community was becoming aware of what has been happening to so many with these shots.

Jeff commented that he sees more and more nurses waking up and beginning to talk about the adverse events and deaths they are witnessing. But when he tries to broach the subject with doctors, "They just look at you like you have spiders coming out of your ears," he said. 'They absolutely cannot deal with it at all—they're totally shut down."

But, remarkably, all over the world, truth is emerging, and justice is being served. In Sweden and Italy, manslaughter, even first-degree murder charges have been brought forth by state prosecutors, against doctors and nurses who gave the actual injections—injections that killed people. One case, the case of Nicholas Sundgren, thirteen, from Gothenburg, who died a horrific death following two Moderna shots, has made the cover of four major newspapers in Sweden, in the spring of 2024.[1]

But there was no autopsy for Tori's baby, Naomi. No way to know exactly what had caused her fateful condition. Autopsies for newborns, by design, are prohibitively expensive—in the tens of thousands—and few can afford it. The only way to prove Naomi's death was caused by the vaccine is by special staining techniques of the placenta and tissue at autopsy—which is very costly and not widely available. Given the constellation of findings, it is clear this was a COVID-19 vaccine-related death.

"I think overall, a lot of people are starting to get it," Jeff says. "And it's because they're seeing a lot of deaths. We knew a boy . . . he was probably around eighteen or nineteen. He was getting ready to ship out for the military, and the following day just dropped dead at school of a heart attack. His dad was . . . the human resource officer at the school. That's not normal—it just does not happen. We know what caused it. People are starting to see it, no doubt about it. We're sharing the dickens out of the documentary [*Shot Dead*]."

But a taboo persists. The White family was not permitted to show the film at their church, as it was deemed too "controversial."

## Unbridled Courage—Sharing Their Story

Tori feels that she is still not out of the woods, but her trauma motivates her to warn others. "I still have nightmares about what I went through with Naomi. It's made me very cautious, this time around, with my son. I don't ever again want to go through . . . what I went through with her, and I don't want to see anybody else [live] that horror. That's why we talk about it."

Her father adds: "From the first moment when [film maker] Teryn Gregson contacted us, we just decided we will re-live it as many times as we have to if it might help somebody out there." Jeff has become a devoted activist, traveling to his state capital, reaching out to his senators, and trying to arrange screenings of *Shot Dead*, wherever he can.

"Here in West Virginia, we've become a part of the medical freedom organization," says Jeff. "This past February, we had a day at the state capitol, lobby day. We got to talk to a couple of our senators and the attorney general who's now . . . running for governor. We feel it's important to be involved in the legislative process to make sure that our politicians know that we're not going to put people through this again. We will not do this again. The people I'm in contact with—they totally agree. We're not going to do this again. That's the line right there; that's the, that's the important line for every American: We're not going thru this again. Because people are talking about . . . they're launching the next one now: 'Bird flu." 'Disease X . . .'

"It's up to us citizens now, . . . to say absolutely not."

Tori rocks back and forth in her rocking chair, and readjusts the blanket around Ares Ryker James White, who reaches out a small arm, with a sleepy stretch, then pulls it back.

"There's still times," she says, "where I'm still kind of . . . scared. Like, the vaccine could maybe have caused some harm to him. Of course, with Naomi, she had the short umbilical cord. But he actually ended up . . . having the same thing. And I've noticed, I guess, a little bit of some of the stuff that happened with Naomi are recurring . . .

"I just want to keep getting Naomi's story out," Tori says in a quiet voice, that also contains deep conviction.

Her father's face softens, as he looks at his daughter and sleeping grandson.

"She's my baby girl.... I think she understands now that her job is to take care of Ares and my job is to fight for him, to fight for the memory of Naomi, and just do what I can in my little corner of the world."

## Modern Day Semmelweis-ian Heroes

The bravery of this young mother and her family have earned my deepest respect and my prayers are extended to all the families who have endured the loss of their loved ones in this catastrophe.

That deep well of respect extends to all the champions of truth—our modern day Semmelweis-ian heroes—the doctors, nurses, scientists, and whistleblowers who include the likes of: Dr. Peter McCullough, Dr. Peter Breggin, Dr. Naomi Wolf, Dr. Vladimir Zelenko (RIP), Dr. Rashid Buttar (RIP), Dr. Arne Burkhardt (RIP), Dr. Luc Montagnier (RIP), Dr. Ben Marble, Dr. Paul Alexander, Dr. Ryan Cole, Dr. Christiane Northrup, Dr. Lee Vliet, Dr. Pierre Kory, Dr. Paul Marik, Dr. Richard Urso, Dr. Mary Talley Bowden, Dr. Meryl Nass, Dr. Stella Emmanuel, Dr. Lee Merritt, Dr. Judy Mikovits, Dr. David Brownstein, Dr. Harvey Risch, Dr. Danial Nagase, Dr. Mel Bruchet, Dr. William Makis, Dr. Bryan Ardis, Dr. Dan McDyer, Dr. Kimberly Biss, Dr. Poppy Daniels, Dr. Richard Bartlett, Jon Fleetwood, Dr. Henry Ealy, Dr. Avery Jackson, Dr. John Witcher, Dr. Stewart Tankersley, Dr. Jordan Vaughn, Kirstin Cosgrove, Dr. Steve Hatfill, Dr. Claire Rogers, Dr. Drew Pinsky, Dr. Kelly Victory, Dr. Eric Feintuch, Robert F. Kennedy Jr., Maggie Thorp, Megan Redshaw, Jack Wolfson, Dr. David Martin, Michelle Spencer, Tiffany Parotto, Sue Peters, Maureen McDonnel, Albert Benevides, John Campbell, Steve Kirsch, Denis Rancourt, Sucharit Bhakdi, and so many others that deserve recognition.

These people have inspired my very difficult journey in speaking truth to power. They have worked tirelessly, often under threat, censorship, loss of jobs, etc., to inform a very unsuspecting public of the mass lies and propaganda that is the Covid psyop. And although, the criminals won't stop, as stated by Tori's dad:

> We have to remain alert, because we can be sure they will be launching the next one... Bird flu. Disease X... We're not done with this battle. You can be sure they have their vaccines all lined up for whatever disease, and whatever fear porn they plan to impose on the public. Now, it's up to each one of us, to not fall for the "script," to be skeptical of "authority."

# The Turning Tide

No one understand this better than Holocaust survivor Vera Sharav. As a six-year-old, her refusal to follow orders literally saved her life. In an October 12, 2020, interview with *Stand for Health Freedom*, Sharav warns:

> There are crossroads in life, where you **have to** make choices. And if you don't, someone who will make the choice for you, is **not** going to make it in your best interest.

We must stay vigilant. And, by the Grace of God, and with every fiber of my Being, I wholeheartedly believe, we will win this fight.

Every day that we resist the lie, we are already—standing together—triumphant.

# Epilogue

*Silence in the face of evil is itself evil.*
—Dietrich Bonhoeffer

We want readers to know that slowly, consistently, the Truth is emerging. Since the time of this writing, there has been an ongoing avalanche of data—from wide-ranging sources and geographic locations—that document the lies and deceptions of the Covid era, along with actions being taken in response to those lies. We include just a few examples below:

- The greatest ray of hope in the last four years, the Czech Republic—a former communist nation—exemplifies a commitment to truth, transparency, and freedom, by publicly and legally releasing more than ten million of their citizens' de-identified health data relevant to the COVID-19 vaccines. The *results irrefutably prove the immense toll of injuries and deaths from COVID-19 vaccines.* Had the United States done this in the first six months of 2021, hundreds of millions would have been saved from death and injuries.
- Children's Health Defense references a study by Denis Rancourt, Joseph Hick, and Christian Linard.[1] with this explosive headline: "Largest Study of its Kind Finds *Excess Deaths During Pandemic Caused by Public Health Response, Not Virus.*"
- As early as September 2023, "*Less than 4 percent of eligible people have gotten updated Covid booster shots, one month into the rollout.*"[2] This seems to indicate that vast majority of Americans do not trust the government's false narrative that the COVID-19 vaccines are safe and effective. In the first quarter of 2024 the COVID-19 vaccine uptake is only 1 percent.[3]

- In a Senate hearing July 11th, 2024, *former CDC Director Robert Redfield testified that mRNA COVID-19 vaccines are "toxic" and should not have been mandated.* He also called for a pause on gain-of-function research.
- The latest Rasmussen Reports national telephone and online survey finds that *53 percent of American adults believe it is likely that side effects of COVID-19 vaccines have caused a significant number of unexplained deaths.*[4]
- As of January 2023, Rasmussen Polls document that *more than 25 percent think they know someone who died from COVID-19 Vaccines.*[5]
- "Ninth Circuit Court Rules Correctly, *COVID-19 mRNA Injections Are Not Legitimate State Interest Due to Being a Treatment, Not a Preventative* | This Is All We Need To DESTROY WHO & FDA"[6]
- A major reason for hope and optimism is that *five states will be suing Pfizer.*[7] These states are: Texas, Utah, Kansas, Mississippi, and Louisiana. In his June 25, 2024, post on X, RFK Jr. points out "That's 10 percent of US States. The tide is turning."
- *European Union countries have destroyed 4 billion Euros worth of COVID vaccines*; including about 200 million unwanted vaccines.[8]
- *Japan is now openly discussing severe vaccine side effects.* Professor Mihazawa of Kyoto University is speaking out aggressively against the COVID-19 vaccines.[9]
- The 5th District Federal Court of Appeals *ruled in favor of the American Association of Physicians and Surgeons, resurrecting freedom of speech.*[10]
- Regulators in the United Kingdom decided against the use of recommending Covid vaccines in pregnant women.[11]

# Endnotes

## Introduction
1. https://www.usaspending.gov/disaster/covid-19?section=award_spending_by_agency.
2. Dr. Luke McLindon, "Totality of Evidence," August 30, 2022, https://totalityofevidence.com/dr-luke-mclindon/.
3. Dr. David E. Martin. "Exposing Moderna; The Star of Plandemic: Indoctornation Reveals the Truth," *London Reel*, September 15, 2020, https://podcasts.apple.com/us/podcast/dr-david-e-martin-exposing-moderna-the-star-of/id474722933?i=1000491311131.

## Chapter 2
1. "Nurse blows whistle on alarming increase in fetal deaths after COVID jab rollout," LifeSiteNews, December 6, 2022, https://www.lifesitenews.com/news/nurse-blows-whistle-on-alarming-increase-in-fetal-deaths-after-covid-jab-rollout/. (accessed 8/28/2024).
2. BNT162b2 5.3.6. "Cumulative Analysis of Post-authorization Adverse Event Reports." https://phmpt.org/document/5–3-6-postmarketing-experience-pdf/. (accessed 8/28/2024).
3. Sage's Newsletter. https://sagehana.substack.com/archive?sort=new.
4. Celia Farber. "Have We Ever Stopped to Think about the Staff on Delivery Wards? This May Be the Hardest Piece You Have Read Yet," *The Truth Barrier* Substack, December 29, 2022, https://celiafarber.substack.com/p/have-we-ever-stopped-to-think-about?utm_source=publication-search.

## Chapter 3
1. Celia Farber. "The Machine Model Of Biology, Denial of The Mystery, Biological Reductionism, And The Scientist Who Tried To Warn Us: Interview With Richard Strohman," *The Truth Barrier* Substack, June 20, 2021, https://celiafarber.substack.com/p/the-machine-model-of-biology-denial?utm_source=publication-search.
2. Banoun H. "mRNA: Vaccine or Gene Therapy? The Safety Regulatory Issues," *Int J Mol Sci.*, 2023 Jun 22;24(13):10514. doi: 10.3390/ijms241310514. PMID: 37445690; PMCID: PMC10342157. https://pubmed.ncbi.nlm.nih.gov/37445690/. (accessed 8/28/2024).
3. Meir Rinde. "The Death of Jesse Gelsinger, 20 Years Later. Gene editing promises to revolutionize medicine. But how safe is safe enough for the patients testing these

therapies?" June 4, 2019, in Science History Institute Museum & Library. https://science history.org/stories/magazine/the-death-of-jesse-gelsinger-20-years-later/ Accessed 7/7/2024.

4. Schädlich A, Hoffmann S, Mueller T, Caysa H, et al. "Accumulation of nanocarriers in the ovary: A neglected toxicity risk?" *Journal of Controlled Release*, Volume 160, Issue 1, May 30, 2021, 105–112. www.sciencedirect.com/science/article/abs/pii/S0168365912000892.

5. Paul Alexander. "Japanese Biodistribution study that Dr. Byram Bridle warned about showing that lipid-nano particles, the mRNA & thus possible spike bio-accumulated in distant tissue & organs based on rat model; Table." Substack, November 9, 2022. https://palexander.substack.com/p/japanese-biodistribution-study-that (accessed 8/28/2024)

6. Maggie Thorp JD and Jim Thorp MD. "The Government Cartel paid billions to Walgreens and CVS not to fill Ivermectin – the question is why. America Out Loud News Monday," May 20, 2024. www.americaoutloud.news/the-government-cartel-paid-billions-to-walgreens-and-cvs-not-to-fill-ivermectin-the-question-is-why/.

7. Dr. Naomi Wolf, "A 360 Degree Attack On Human Reproduction From The Pfizer Docs," Real America's Voice, *The Charlie Kirk Show*, https://rumble.com/v1glp8t-dr.-naomi-wolf-a-360-degree-attack-on-human-reproduction-from-the-pfizer-do.html (Accessed 8/28/2024).

8. Maggie Thorp JD, Jim Thorp MD. "A call for Immediate Moratorium on the use of COVID-19 Vaccines in pregnant women," March 5, 2024, *America Out Loud*, www.americaoutloud.news/a-call-for-immediate-moratorium-on-the-use-of-covid-19-vaccines-in-pregnant-women/.

9. Lin X, Botros B, Hanna M, Gurzenda E, Manzano De Mejia C, Chavez M, Hanna N. "Transplacental transmission of the COVID-19 vaccine messenger RNA: evidence from placental, maternal and cord blood analyses postvaccination," *American Journal of Obstetrics and Gynecology* (2024), doi: https://doi.org/10.1016/j.ajog.2024.01.022.

10. Maggie Thorp JD, Jim Thorp MD. "A call for Immediate Moratorium on the use of COVID-19 Vaccines in pregnant women," *America Out Loud*, March 3, 2024, www.americaoutloud.news/a-call-for-immediate-moratorium-on-the-use-of-covid-19-vaccines-in-pregnant-women/.

11. Ibid.

12. Ibid.

13. Lin X, Botros B, Hanna M, et al. "Transplacental transmission of the COVID-19 vaccine messenger RNA: evidence from placental, maternal, and cord blood analyses post-vaccination," *Am J Obstet Gynecol*, Jan 31, 2024, Vol. 230 (6), www.ajog.org/article/S0002–9378(24)00063–2/fulltext (accessed 8/28/2024).

14. Ibid.

15. Mehra, M., Desai, S. et al. "Retracted: Hydroxychloroquine or chloroquine with or without a macrolide for treatment of COVID-19: a multinational registry analysis," *Lancet*, May 22, 2020, https://www.thelancet.com/article/S0140–6736(20)31180–6/fulltext.

16. Shimabukuro, T., Kim, S. et al. "Preliminary Findings of mRNA Covid-19 Vaccine Safety in Pregnant Persons," *New England Journal of Medicine*, April 21, 2021, https://www.nejm.org/doi/full/10.1056/NEJMoa2104983.

17. Aldén M, Olofsson Falla F, Yang D, Barghouth M, Luan C, Rasmussen M, De Marinis Y. "Intracellular Reverse Transcription of Pfizer BioNTech COVID-19 mRNA Vaccine BNT162b2 In Vitro in Human Liver Cell Line," *Curr Issues Mol Biol.*, 2022

Endnotes 171

Feb 25;44(3):1115–1126. doi: 10.3390/cimb44030073. PMID: 35723296; PMCID: PMC8946961.

[18] Hanna N, Heffes-Doon A, Lin X, Manzano De Mejia C, Botros B, Gurzenda E, Nayak A. "Detection of Messenger RNA COVID-19 Vaccines in Human Breast Milk," *JAMA Pediatr.*, 2022 Dec 1;176(12):1268–1270. doi: 10.1001/jamapediatrics.2022.3581. Erratum in: *JAMA Pediatr.* 2022 Nov 1;176(11):1154. doi: 10.1001/jamapediatrics .2022.4568. PMID: 36156636; PMCID: PMC9513706. https://pubmed.ncbi.nlm .nih.gov/36156636/.

[19] Hanna N, De Mejia CM, Heffes-Doon A, et al. "Biodistribution of mRNA COVID-19 vaccines in human breast milk," *eBioMedicine Part of The Lancet,* Discovery Science. October 1993, Volume 96, 104800. www.thelancet.com/journals/ebiom/article/PIIS2352 -3964(23)00366-3/fulltext (Accessed 8/28/2024).

[20] Mikolaj Raszek. "Vaccine Genome Insertion," January 2024, www.youtube.com/watch?v =y7dMTA-fD5Q (Accessed 8.28/2024).

# Chapter 5

[1] Jacques Ellul. *Propaganda*: *The Formation of Men's Attitudes* (New York: Alfred A. Knopf, 1965).

[2] DailyClout. "Pfizer Documents Analysis Volunteers' Reports eBook: Find Out What Pfizer, FDA Tried to Conceal." War Room/DailyClout, January 1, 2023.

[3] Wikipedia Direct-to-consumer advertising (DTCA). https://en.wikipedia.org/wiki /Direct-to-consumer_advertising.

[4] Beth Snyder Bulik. "HHS rolls $250M ad campaign with Fauci,m science focus front and center." Fierce Pharma, December 14, 2020. www.fiercepharma.com/marketing /hhs-rolls-250-million-ad-campaign-fauci-and-science-focus-front-and-center.

[5] Maggie Thorp JD and Jim Thorp MD. "Tentacles of a Covert and Exploitative Propaganda Machine Compliments of the US Government," *America Out Loud News*, October 28, 2022, www.americaoutloud.news/tentacles-of-a-covert-and-exploitative-propaganda -machine-compliments-of-the-us-government/ (accessed 8/28/2024).

[6] Ibid.

[7] Ibid.

[8] "U.S. Department of Health and Human Services Launches Nationwide Network of Trusted Voices to Encourage Vaccination in Next Phase of COVID-19 Public Education Campaign," Wayback Machine, April 1, 2021. Captured by screenshot on internet archive Wayback Machine. https://web.archive.org/web/20230601074707/https://web .archive.org/web/20210401225102/https://www.hhs.gov/about/news/2021/04/01/hhs -launches-nationwide-network-trusted-voices-encourage-vaccination-next -phase-covid-19-public-education-campaign.html.

[9] USA Spending.gov. https://www.usaspending.gov/search/?hash=2b9bbf7349e6c520a551 64cbe34c6321; Maggie Thorp JD and Jim Thorp MD. "FOIA Reveals Troubling Relationship between HHS/CDC & the American College of Obstetricians and Gynecologists," *America Out Loud News*, May 7, 2023. https://www.americaoutloud.news /foia-reveals-troubling-relationship-between-hhs-cdc-the-american-college-of-obstetricians -and-gynecologists/.

[10] Maggie Thorp JD and Jim Thorp MD. "FOIA Reveals Troubling Relationship between HHS/CDC & the American College of Obstetricians and Gynecologists," *America Out

Loud News. May 7, 2023, https://www.americaoutloud.news/foia-reveals-troubling-relationship-between-hhs-cdc-the-american-college-of-obstetricians-and-gynecologists/.

[11] See webpage maintained by: American College of Obstetricians and Gynecologists (ACOG). 2023. "COVID-19 Vaccines and Pregnancy: Conversation Guide—Key Recommendations and Messaging for Clinicians." Accessed July/14/2024. www.acog.org/covid-19/covid-19-vaccines-and-pregnancy-conversation-guide-for-clinicians.

[12] Ibid.

[13] ProPublica Nonprofit Explorer. American College of Obstetricians and Gynecologists. https://projects.propublica.org/nonprofits/organizations/900489809 (accessed 8/28/2024).

[14] Alicia Parlapiano et al. "Where $5 Trillion in Pandemic Stimulus Money Went." *New York Times*, https://www.nytimes.com/interactive/2022/03/11/us/how-covid-stimulus-money-was-spent.html. March 11, 2022. (Accessed 10/12/2024).

[15] Centers for Disease Control and Prevention. 2023. "HHS Provider Relief Fund." Data. https://data.cdc.gov/Administrative/HHS-Provider-Relief-Fund/kh8y-3es6/data_preview.

[16] Maggie Thorp JD and Jim Thorp MD. "COVID-10 Government Relief Funds Turned the Healthcare Industry on Its Head," *America Out Loud News*, December 10, 2023. https://www.americaoutloud.news/covid-19-government-relief-funds-turned-the-healthcare-industry-on-its-head/.

[17] National Citizens Inquiry, "CBC–Government Funded Media," Nationalcitizensinquiry.ca, https://nationalcitizensinquiry.ca/wp-content/uploads/2023/07/OT-15-Palmer-Second-Testimony-May-18.pdf.

[18] Dr. Stone (@gurillaapp). "Helpful debunking of shills' myths . . ." X (formerly Twitter), May 16, 2024, https://x.com/gurillaapp/status/1791204725460763124.

[19] Helen Metella. "Campaign countering vaccine misinformation receives new funding of $1.75 million," University of Alberta, April 29, 2021, www.ualberta.ca/en/law/about/news/2021/4/new-funding.html.

[20] Government of Canada. Grants and Contributions. Caulfield, Timothy A. https://search.open.canada.ca/grants/?sort=agreement_value+desc&page=1&search_text=Caulfield+

[21] Maggie Thorp JD, Jim Thorp MD. "COVID-19 Government Relief Funds Turned the Healthcare Industry on Its Head." *America Out Loud*, December 10, 2023, www.americaoutloud.news/covid-19-government-relief-funds-turned-the-healthcare-industry-on-its-head/.

[22] Maggie Thorp JD, Jim Thorp MD. "US Government coerced leaders of faith to push COVID-19 vaccines on Americans." *America Out Loud*, January 14, 2024, www.americaoutloud.news/us-government-coerced-leaders-of-faith-to-push-covid-19-vaccines-on-americans/.

[23] Ibid.

[24] Maggie M Thorp JD. Jim Thorp MD. "Tentacles of a Covert and Exploitative Propaganda Machine Compliments of the US Government," *America Out Loud*, October 28, 2022. www.americaoutloud.news/tentacles-of-a-covert-and-exploitative-propaganda-machine-compliments-of-the-us-government/.

[25] Maggie Thorp JD and Jim Thorp MD. "US Government coerced leaders of faith to push COVID-19 vaccines on Americans," *America Out Loud*, January 14, 2024, www.americaoutloud.news/us-government-coerced-leaders-of-faith-to-push-covid-19-vaccines-on-americans/ (Accessed 8/24/2024).

[26] Ibid.

[27] Ibid.

# Endnotes

28  Ibid.
29  Ibid.
30  Ibid.
31  *See full Campaign background: https://www.covid.gov/be-informed/campaign-background; *Read about the strategy and approach to reaching Campaign audiences: https://www.covid.gov/get-involved/covid-19-campaign-strategy-and-approach
32  *Learn about the COVID-19 Community Corps: https://www.covid.gov/get-involved/community-corps.*
33  Pickworth, Carin. "I'm in Love with my Obstetrician, and I'm not Alone." News.com.au. *KidSpot*, April 5, 2018. Accessed May 2, 2023. www.kidspot.com.au/pregnancy/labour/im-in-love-with-my-obstetrician-and-im-not-alone/news-story/1fc5007077f517444c29fe53acecce56.
34  Ibid.
35  WCA_FemalePowerhouse_Infographic_2018.pdf. https://womenschoiceaward.com/wp-content/uploads/2018/01/WCA_FemalePowerhouse_Infographic_2018.pdf.
36  WCA_FemalePowerhouse_Infographic_2018.pdf., p.8. https://womenschoiceaward.com/wp-content/uploads/2018/01/WCA_FemalePowerhouse_Infographic_2018.pdf.
37  Kates, Graham. "Inside the $250 Million Effort to Convince Americans the Coronavirus Vaccines are Safe." CBS News, December 23, 2020. Accessed May 2, 2023. www.cbsnews.com/news/covid-vaccine-safety-250-million-dollar-marketing-campaign/.
38  J. J. Carrell, Brian O'Shea. DailyClout Unleash Liberty. E10S2: "Biden's Border Lies/Jason James Gives Views from Canada," DailyClout, February 5, 2024, https://dailyclout.io/e10s2-bidens-border-lies-jason-james-gives-views-from-canada/ (Accessed 8/24/2024)

## Chapter 6

1  Harold Evans. "Thalidomide: how men who blighted lives of thousands evaded justice." *The Guardian*, November 14, 2014, www.theguardian.com/society/2014/nov/14/-sp-thalidomide-pill-how-evaded-justice. (Accessed 8/28/2024).
2  BBC News, "German thalidomide maker Gruenenthal issues apology," BBC News, September 1, 2021, https://www.bbc.com/news/health-19443910.
3  Harold Evans. "Thalidomide: how men who blighted lives of thousands evaded justice." *The Guardian*, November 14, 2014, www.theguardian.com/society/2014/nov/14/-sp-thalidomide-pill-how-evaded-justice. (Accessed 8/28/2024).
4  Roger Williams, "The Nazis and Thalidomide: The Worst Drug Scandal of All Time," *Newsweek*, September 10, 2012, www.newsweek.com/nazis-and-thalidomide-worst-drug-scandal-all-time-64655.
5  Harold Evans. "Thalidomide: how men who blighted lives of thousands evaded justice." *The Guardian*, November 14, 2014, www.theguardian.com/society/2014/nov/14/-sp-thalidomide-pill-how-evaded-justice. (Accessed 8/28/2024).
6  Marcus Williamson. Frances Oldham Kelsey: "Scientist who blocked the sale of thalidomide in the US, saving countless unborn children from its dire effects," *Independent*, August 16, 2015, https://amp.theguardian.com/society/2016/jan/24/mikey-argy-thalidomide-campaigner-attacking-the-devil-documentary (Accessed 7/7/2024).
7  FDAConsumer Health Information, "Kefauver-Harris Amendments Revolutionized Drug Development," PDF, www.gvsu.edu/cms4/asset/F51281F0–00AF-E25A-5BF

632E8D4A243C7/kefauver-harris_amendments.fda.thalidomide.pdf. (accessed, October 6, 2024).
8   Ibid.
9   Ibid.
10  "HHS Covid-19 and Flu Public Education Campaign," HHS, https://www.covid.gov/sites/default/files/2021–04/WCDT%20Public%20Education%20Campaign%20Backgrounder.pdf.
11  Ibid.
12  Dieckmann W., Davis M., Rynkiewicz L., Pottinger R. "Does the administration of diethylstilbestrol during pregnancy have therapeutic value?" *Am. J. Obstet. Gynecol,* 1953; 66:1062–1081. doi: 10.1016/S0002–9378(16)38617–3.
13  Arthur L. Herbst, "Adenocarcinoma of the Vagina—Association of Maternal Stilbestrol Therapy with Tumor Appearance in Young Women," *The New England Journal of Medicine (NEJM),* April 22, 1971.
14  Loizzo A., Gatti G.L., Macri A., Moretti G., Ortolani E., Palazzesi S. "The case of diethylstilbestrol treated veal contained in homogenized baby-foods in Italy. Methodological and toxicological aspects." *Ann. Ist. Super. Sanita.,* 1984;20:215–220.

## Chapter 7

1   Maggie Fox. "CDC to pregnant women: Get Vaccinated against Covid-19." CNN, September 29, 2021, www.cnn.com/2021/09/29/health/pregnancy-covid-vaccine-cdc-wellness/index.html (Accessed 8/28/2024).
2   Naomi Wolf. "*Facing the Beast: Courage, Faith, and Resistance in a New Dark Age,*" (Vermont: Chelsea Green, 2023), 95.

## Chapter 9

1   X (formerly Twitter) post by James A Thorp MD, August 21, 2023. 127K views. https://x.com/jathorpmfm/status/1693815290407072085.
2   Ibid.
3   Raphael Lataster, PhD. "COVID-19 vaccine negative effectiveness in UK booster study." August 7, 2024. https://okaythennews.substack.com/p/covid-19-vaccine-negative-effectiveness-f14.
4   Meryl Nass, "Negative COVID vaccine efficacy. When the remedy becomes the poison. These vaccines poison in multiple ways, but here I am confining the evidence to the vaccine-caused increase in susceptibility to the infection it was intended to prevent." Meryl Nass Substack, July 6, 2024. https://merylnass.substack.com/p/negative-covid-vaccine-efficacy-when.

## Chapter 10

1   ABOG American Board of Obstetrics & Gynecology, September 27, 2021, www.abog.org/about-abog/news-announcements/2021/09/27/statement-regarding-dissemination-of-covid-19-misinformation.
2   FSMB, Federation of State Medical Boards. www.fsmb.org/ (Accessed 7.11.2024).
3   American Board of Medical Specialties. "ABMS Issues Statement Supporting Role of Medical Professionals in Preventing COVID-19 Misinformation," September 13,

# Endnotes

2021, www.abms.org/newsroom/abms-issues-statement-supporting-role-of-medical-professionals-in-preventing-covid-19-misinformation/ (Accessed 7.11.2024).

4. ABOG, American Board of Obstetrics & Gynecology. "Revocation of Diploma or Certificate," www.abog.org/about-abog/policies/revocation-of-diploma-or-certificate (Accessed July 11, 2024).

5. Kara Grant. "More States Seeing Uptick of Pregnant COVID Patients in ICU's. Nearly all are unvaccinated, sources say." MEDPAGE TODAY, September 15, 2021. [Please note this is a lay-writer and this is on par with a grocery market tabloid]

6. Please Note Well—In the ABOG threat, this link states "CDC Reported" but actually brings you to an article from CNN. Here is the CNN article that this hyperlink directs on to: Maggie Fox and Jamie Gumbrecht. "Vaccine protection against Covid-19 wanes over time, especially for older people, CDC says," CNN, September 23, 2021. (Accessed July 11/2024).

7. ACOG, The American College of Obstetricians and Gynecologists. "ACOG and SMFM Recommend COVID-19 Vaccination for Pregnant Individuals," July 30, 2021. www.acog.org/news/news-releases/2021/07/acog-smfm-recommend-covid-19-vaccination-for-pregnant-individuals (Accessed 7/11/2024).

8. CDC. "COVID-19 Vaccines While Pregnant or Breastfeeding." Updated on March 8, 2024.

9. ABOG About ABOG. www.abog.org/about-abog/about-abog.

10. ABOG, "Statement Regarding Dissemination of COVID-19 Misinformation," www.abog.org/about-abog/news-announcements/2021/09/27/statement-regarding-dissemination-of-covid-19-misinformation.

11. ABOG, American Board of Obstetrics & Gynecology. "Statement Regarding Dissemination of COVID-19 Misinformation," September 27, 2021. www.abog.org/about-abog/news-announcements/2021/09/27/statement-regarding-dissemination-of-covid-19-misinformation (Accessed 8/28/2024)

12. Kara Grant, Enterprise & Investigative Writer, MedPage Today, September 15, 2021.

13. Pineles BL, Goodman KE, Pineles L, O'Hara LM, et al. "Pregnancy and the Risk of In-Hospital Coronavirus Disease 2019 (COVID-19) Mortality," *Obstet Gynecol.*, 2022 May 1;139(5):846–854. doi: 10.1097/AOG.0000000000004744. Epub 2022 Apr 5. PMID: 35576343; PMCID: PMC9015030. www.ncbi.nlm.nih.gov/pmc/articles/PMC9015030/.

14. Kara Grant. "'Alarming' Number of Pregnant Women Admitted to Alabama ICU's—The trend points to just how important it is for pregnant women to get vaccinated, physician's say." *MedPage Today.* August 24, 2021. www.medpagetoday.com/special-reports/exclusives/94193.

15. Kara Grant. 'Alarming' Number of Pregnant Women Admitted to Alabama ICUs." *MedPage Today.* August 24, 2021, www.medpagetoday.com/special-reports/exclusives/94193 (Accessed 8/28/2024)

16. Alicia Rohan. "UAB received more than $325 million in research funding from the National Institutes of Health in 2020," March 9, 2021, www.uab.edu/news/research/item/11887-uab-received-more-than-325-million-in-research-funding-from-the-national-institutes-of-health-in-2020.

17. Ryan McCain. "Heersink School of Medicine receives over $610 million in research funding for 2023," UAB, April 10, 2024, www.uab.edu/medicine/news/latest/item/3017-heersink-school-of-medicine-receives-over-610-million-in-research-funding-for-2023.

[18] Mehra MR, Desai SS, Ruschitzka F et al. "RETRACTED: Hydroxychloroquine or chloroquine with or without a macrolide for treatment of COVID-19: a multinational registry analysis," *Lancet* May 22, 2020, www.thelancet.com/journals/lancet/article/PIIS0140–6736%2820%2931180–6/fulltext (Accessed 8/28/2024).

[19] Isabella Murray. "'COVID kills moms.' 8 pregnant women died of the virus recently, MS health dept. says." *Sun Herald*, September 9, 2021, www.sunherald.com/news/coronavirus/article254077793.html.

[20] Pineles BL, Goodman KE, Pineles L, O'Hara LM; et al. "In-Hospital Mortality in a Cohort of Hospitalized Pregnant and Nonpregnant Patients With COVID-19," *Ann Intern Med.*, 2021 Aug;174(8):1186–1188. doi: 10.7326/M21–0974. Epub 2021 May 11. PMID: 33971101; PMCID: PMC8251936. www.ncbi.nlm.nih.gov/pmc/articles/PMC8251936/ (Accessed 8/28/2024).

[21] See Page 170–171 for discussion of the data manipulation in the Shimabukuro article in: Thorp KE, Thorp MM, Thorp EM, Thorp JA. "COVID-19 & Disaster Capitalism—Part I." *G Med Sci.*, 2022; 3(1):159–178. www.doi.org/10.46766/thegms.medethics.22071901.

[22] Thorp MM, Thorp JA. "Pushing COVID-19 Shots in Pregnancy: The Greatest Ethical Breach in the History of Medicine," *America Out Loud News*, February 12, 2023. www.americaoutloud.news/pushing-covid-19-shots-in-pregnancy-the-greatest-ethical-breach-in-the-history-of-medicine/ (Accessed 7/11/2024).

[23] Thorp, J. A., Benavides A., Thorp, M. M., McDyer, D. C., Biss, K. O., Threet, J. A., & McCullough, P. A., "Are COVID-19 Vaccines in Pregnancy as Safe and Effective as the U.S. Government, Medical Organizations, and Pharmaceutical Industry Claim? Part II." Preprints 2024, 2024070069. https://doi.org/10.20944/preprints202407.0069.v1 www.preprints.org/manuscript/202407.0069/v1.

[24] Thorp, J. A., Benavides A., Thorp, M. M., McDyer, D. C., Biss, K. O., Threet, J. A., & McCullough, P. A., (2024). "Are COVID-19 Vaccines in Pregnancy as Safe and Effective as the U.S. Government, Medical Organizations, and Pharmaceutical Industry Claim? Part I." Preprints. https://doi.org/10.20944/preprints202406.2062.v1

[25] Thorp, J. A., Benavides, A., Thorp, M. M., McDyer, D. C., Biss, K. O., Threet, J. A., & McCullough, P. A. (2024). "Are COVID-19 Vaccines in Pregnancy as Safe and Effective as the U.S. Government, Medical Organizations, and Pharmaceutical Industry Claim? Part II." Preprints. https://doi.org/10.20944/preprints202407.0069.v1

[26] Maggie Fox and Jamie Gumbrecht. "Vaccine protection against Covid-19 wanes over time, especially for older people, CDC says," CNN, September 23, 2021, www.cnn.com/2021/09/22/health/cdc-vaccine-advisers-booster-wane/index.html (Accessed 8/28/2024).

# Chapter 11

[1] REACT19. "Science-based support for people suffering from long-term COVID-19 vaccine effects." https://react19.org/ Accessed July 11, 2024

[2] James A Thorp MD. "Open Letter to The American Board of Obstetrics & Gynecology (ABOG)," January 12, 2022, www.rodefshalom613.org/wp-content/uploads/2022/01/Thorp-ABOG-Letter-.01.12.2021.pdf

[3] Ibid.

[4] Aldén M, Olofsson Falla F, Yang D, Barghouth M, Luan C, Rasmussen M, De Marinis Y. "Intracellular Reverse Transcription of Pfizer BioNTech COVID-19 mRNA

Vaccine BNT162b2 In Vitro in Human Liver Cell Line," *Curr Issues Mol Biol*, 2022 Feb 25;44(3):1115–1126. doi: 10.3390/cimb44030073. PMID: 35723296; PMCID: PMC8946961. *www.ncbi.nlm.nih.gov/pmc/articles/PMC8946961/* .

5   Hanna N, Heffes-Doon A, Lin X, Manzano De Mejia C, Botros B, Gurzenda E, Nayak A. "Detection of Messenger RNA COVID-19 Vaccines in Human Breast Milk," *JAMA Pediatr.*, 2022 Dec 1;176(12):1268–1270. doi: 10.1001/jamapediatrics.2022.3581. Erratum in: JAMA Pediatr. 2022 Nov 1;176(11):1154. doi: 10.1001/jamapediatrics.2022.4568. PMID: 36156636; PMCID: PMC9513706. *https://pubmed.ncbi.nlm.nih.gov/36156636/*.

6   Hanna N, De Mejia CM, Heffes-Doon, Lin X, et al. "Biodistribution of mRNA COVID-19 Vaccines in human breast milk." *eBioMedicine/The Lancet*, September 19, 2023, 96(104800), https://doi.org/10.1016/j.ebiom.2023.104800. www.thelancet.com/journals/ebiom/article/PIIS2352-3964(23)00366-3/fulltext.

# Chapter 12

1   "Dr. Peter McCullough Testifies Before Texas Senate Health & Human Services Committee." June 27, 2022. https://rumble.com/v1aei2z-dr.-peter-mccullough-testifies-before-texas-senate-health-and-human-service.html.

2   Mello.B33 (@B33Mello). 2022. "8/While the WHO changed the definition of 'pandemic' just before the H1N1 outbreak in 2009, last year, the CDC downgraded the definition of 'vaccine' from a product that produces immunity to a preparation that stimulates an immune response for protection." X (formerly Twitter), April 17, 2022. https://x.com/B33Mello/status/1515708147842879491.

3   Ibid.

4   Thomas Massie @RepThomasMassie. 2021. "Check out @CDCgov's evolving definition of 'vaccination.' They've been busy at the Ministry of Truth," X (formerly Twitter), September 8, 2021, https://x.com/RepThomasMassie/status/1435606845926871041.

5   Katie Camero. "Why did CDC change its definition for 'vaccine'? Agency explains move as skeptics lurk," *Miami Herald*, September 27, 2021, www.miamiherald.com/news/coronavirus/article254111268.html#storylink=cpy.

6   Brook Jackson, @IamBrookJackson. "The U.S. military definition of vaccine hasn't changed," X (formerly Twitter), September 22, 2023, https://x.com/IamBrookJackson/status/1705387936735510850.

7   Celia Farber, "Ten Fatal Errors: Scientists Attack Paper That Established Global PCR Driven Lockdown," UncoverDC, December 3, 2023, https://uncoverdc.com/2020/12/03/ten-fatal-errors-scientists-attack-paper-that-established-global-pcr-driven-lockdown.

8   Ezekial J. Emanuel, Céline Gounder et al. "Take whatever COVID vaccine you can get. All of them stop death and hospitalization," *USA Today*, February 12, 2021, https://www.usatoday.com/story/opinion/2021/02/12/all-covid-vaccines-stop-death-severe-illness-column/6709455002/.

9   Thorp KE, Thorp JA, Thorp EM. "COVID-19 and the Unraveling of Experimental Medicine - Part III." *G Med Sci.*, 2022; 3(1):118–158. www.doi.org/10.46766/thegms.pubheal.22042302.

10  The website for this data is referenced here and has also been downloaded for permanent archive: "Study NCT04754594 to Evaluate the Safety, Tolerability, and Immunogenicity of BNT162b2 Against COVID-19 in Healthy Pregnant Women 18 Years of Age

and Older." Submitted Date: July 14, 2023 (v21). https://classic.clinicaltrials.gov/ct2/history/NCT04754594?V_21&embedded=true.

11. Children's Health Defense. (2022.) *Turtles All The Way Down: Vaccine Science and Myth*. Eds. Zoey O'Toole and Mary Holland.

12. Pineles BL, Goodman KE, Pineles L, O'Hara LM, Nadimpalli G, Magder LS, Baghdadi JD, Parchem JG, Harris AD. "In-Hospital Mortality in a Cohort of Hospitalized Pregnant and Nonpregnant Patients With COVID-19," *Ann Intern Med.*, 2021 Aug;174(8):1186–1188. doi: 10.7326/M21–0974. Epub 2021 May 11. PMID: 33971101; PMCID: PMC8251936. https://pubmed.ncbi.nlm.nih.gov/33971101/.

13. Thorp JA, Rogers C; Deskevich, MP, Tankersley S, Benavides A, Redshaw, M.D.; McCullough, P.A. "COVID-19 Vaccines: The Impact on Pregnancy Outcomes and Menstrual Function," *Journal of the American Physicians & Surgeons*, Spring 2023; 28(1) (see Figure 4, p. 32.) www.jpands.org/vol28no1/thorp.pdf.

14. Blaylock RL, Faria M. "New concepts in the development of schizophrenia, autism spectrum disorders, and degenerative brain diseases based on chronic inflammation: A working hypothesis from continued advances in neuroscience research." *Surg Neurol Int.*, 2021 Nov 8;12:556. doi: 10.25259/SNI_1007_2021. PMID: 34877042; PMCID: PMC8645502.

15. Blaylock, RL. "Covid-19 pandemic: What is the truth?" *Surg Neurol Int.*, 2021 Dec 8; 12:591. doi: 10.25259/SNI_1008_2021.

16. Pfizer Documents Investigation Team, Amy Kelly, Daly Clout LLC. War Room / DailyClout Pfizer Documents Analysis Volunteer's Reports eBook: Find Out What Pfizer, FDA Tried to Conceal. www.amazon.com/DailyClout-Documents-Analysis-Volunteers-Reports-ebook/dp/B0BSK6LV5D

17. Brenda Baletti. "Exclusive: Healthy 32-Year-Old Given 3 days to Live After Pfizer Shots Led to Rare Autoimmune Disorder," October 4, 2023, https://childrenshealthdefense.org/defender/cody-hudson-pfizer-covid-vaccine-injury/.

18. Personal communication with Heather Hudson. Heather Hudson is a medical researcher and authored two case reports involving COVID-19 vaccine injury and fatality.

19. Ibid.

20. Balbona EJ. "A case of COVID mRNA Vaccine Linked Antiphospholipid Syndrome." ISSN:2835–7914, March 28, 2023. a50a8933-f412–446e-bc78–287083a87432 [Personal Communication]

21. Naomi Wolf. *The Bodies of Others: The New Authoritarians, COVID-19 and the War Against the Human* (Livermore: All Seasons Press, 2022).

22. Jian-Pei Huang et al. "Nanoparticles can cross mouse placenta and induce trophoblast apoptosis," PubMed, December 2015.

23. Bannon's War Room on Rumble. "Naomi Wolf Lays Out The Case For Banning Covid Vaccines For Pregnant Women," DailyClout, (video), https://dailyclout.io/naomi-wolf-lays-out-the-case-for-banning-covid-vaccines-for-pregnant-women/.

24. Naomi Wolf. *Facing the Beast: Courage, Faith and Resistance in a New Dark Age* (Vermont: Chelsea Green 2023), 96.

25. Doctors & Scientists with Brian Hooker, PhD. "CATEGORY X with Special Guest Kimberly Biss, MD." https://live.childrenshealthdefense.org/chd-tv/shows/good-morning-chd/category-x-with-special-guest-kimberly-biss-md/

26. Parotto T, Thorp JA, Hooker B, Mills PJ, Newman J, Murphy L, et al. "COVID-19 and the surge in Decidual Cast Shedding," *G Med Sci.*, 2022; 3(1): 107- 117. www.doi.org/10.46766/thegms.pubheal.22041401.

27   Ibid.
28   Ibid.
29   Ibid.
30   BNT162b2 5.3.6. "Cumulative Analysis of post-authorization Adverse Event Reports," https://phmpt.org/wp-content/uploads/2022/04/reissue_5.3.6-postmarketing-experience.pdf (Accessed 6/25/2024).
31   Mathew Aldred. "Kevin McKernan on Vaccine DNA Contamination. The regulators could have done these tests cheaply, but instead they just trusted the word of their Big Pharma sponsors," February 24, 2024. https://mathewaldred.substack.com/p/kevin-mckernan-on-vaccine-dna-contamination.
32   World Council for Health. "WCH Expert Panel Finds Cancer-Promoting DNA Contamination in COVID-19 Vaccines." October 10, 2023. https://worldcouncilforhealth.org/news/news-releases/dna-contamination-covid-19-vaccines/ (Accessed 7/12/2024).
33   Maggie Thorp JD and Jim Thorp MD. "A call for Immediate Moratorium on the use of COVID-19 Vaccines in pregnant women," America Out Loud, March 3, 2024. www.americaoutloud.news/a-call-for-immediate-moratorium-on-the-use-of-covid-19-vaccines-in-pregnant-women/ (Accessed 7/12/2024).
34   Lin X, Botros B, Hanna M, Gurzenda E, De Mejia CM, Chavez M, Hanna N. "Transplacental transmission of the COVID-19 vaccine messenger RNA: evidence from placental, maternal, and cord blood analyses postvaccination," *Am J Obstet Gynecol.*, 2024 Jun;230(6):e113-e116. doi: 10.1016/j.ajog.2024.01.022. Epub 2024 Feb 1. PMID: 38307473. https://pubmed.ncbi.nlm.nih.gov/38307473/.
35   Lioness of Judah Ministry. Breaking: "Pfizer's Jab Contains the SV40 Sequence Which Is Known as a Promoter of the Cancer Virus. Japanese professor Murakami of Tokyo University of Science makes an astonishing finding." Substack, April 30, 2023. https://lionessofjudah.substack.com/p/breaking-pfizers-jab-contains-the (Accessed Date 7/12/2024).
36   Ignaz Semmelweis. *The Etiology, the Concept and the Prophylaxis of Childbed Fever together with the "Open Letters,"* privately printed for the Members of the Classics of Obstetrics & Gynecology Library. Translated by Frank P. Murphy MD. Edited with Commentary and Translation of the Semmelweis "Open Letters" by Sherwin B. Nuland, MD and Ferenc A. Gyorgyey. Birmingham, 1990. Copy number 996 of a Limited Edition of 3,000 copies in the Library of James Alan Thorp, MD.
37   Wikipedia. "Ignaz Semmelweis. Last modified September 23, 2024. https://en.wikipedia.org/w/index.php?title=Ignaz_Semmelweis&action=history.

# Chapter 13

1   Celia Farber. "Where To Go To Follow The Unfolding Moderna Shot Manslaughter Case In Sweden, And A Few Words About Stepped Up Engagement In The War." "Nothing Is More Fun Than Fighting Evil.," -Barry Farber. June 30, 2024. https://celiafarber.substack.com/p/where-to-go-to-follow-the-unfolding (Accessed 7/12/2024).

# Epilogue

1   Denis Rancourt, Joseph Hickey, and Christian Linard, "Spatiotemporal variation of the excess all-cause mortality in the world (125 countries) during the Covid period 2020–

2023 regarding socio-economic factors and public-health and medical interventions," July 19, 2024, https://correlation-canada.org/covid-excess-mortality-125-countries/?utm_source=substack&utm_medium=email.

2. Aria Bendix. "Less than 4% of eligible people have gotten updated Covid booster shots, one month into the rollout," NBC News, September 23, 2022, www.nbcnews.com/health/health-news/updated-covid-booster-shots-doses-administered-cdc-rcna48960 (Accessed 7/7/2024).

3. Personal Communication from Dr. Peter A. McCullough, MD, MPH.

4. Rasmussen Reports. "More Than Half Suspect COVID-19 Vaccines Have Caused Deaths," January 12, 2024, www.rasmussenreports.com/public_content/politics/public_surveys/more_than_half_suspect_covid_19_vaccines_have_caused_deathsl (Accessed 7/7/2024).

5. Rasmussen Reports. "'Died Suddenly'? More Than 1–4 Think Someone They Know Died From COVID-19 Vaccines," Rasmussen Reports, January 2, 2023. www.rasmussenreports.com/public_content/politics/public_surveys/died_suddenly_more_than_1_in_4_think_someone_they_know_died_from_covid_19_vaccines. (Accessed 7/7/2024)

6. Interest of Justice. "Ninth Circuit Court Rules Correctly COVID-19 mRNA Injections Are Not Legitimate State Interest Due to Being a Treatment, Not a Preventative | This Is All We Need to DESTROY WHO & FDA," June 8, 2024, https://interestofjustice.substack.com/p/ninth-circuit-court-rules-correctly (Accessed 7/7/2024).

7. Peter Imanuelsen. "The Freedom Corner with PeterSweden. HUGE: Five states will be SUING Pfizer. Kansas and Idaho among the states to be suing Pfizer over misleading claims about their covid injections," Substack, June 24, 2024, https://petersweden.substack.com/p/huge-five-states-will-be-suing-pfizer (Accessed 7/7/2024).

8. Carlo Martuscelli and Hanne Cokelaere. "EU countries destroy €4B worth of COVID vaccines. A Politico analysis shows that more than 200M unwanted coronavirus jabs have been dumped," *Politico*, December 18, 2023 (Accessed 7/7/2024).

9. Lioness of Judah Ministry. "Exposing The Darkness: Open Discussion About the Harms of the COVID Injections on Japanese TV," Substack, June 1, 2024. https://lionessofjudah.substack.com/p/open-discussion-about-the-harms-of (Accessed 7/7/2024).

10. Maggie Thorp JD, Jim Thorp MD. "The war on free speech! Medical boards accused of collusion with feds as 'State Actor' status looms," June 12, 2024, www.americaoutloud.news/the-war-on-free-speech-medical-boards-accused-of-collusion-with-feds-as-state-actor-status-looms/ (Accessed 7/7/2024).

11. Brenda Baletti. "UK Regulators Decide Pregnant Women Don't Need COVID Vaccines." *The Defender*. Children's Health Defense News & Views. November 20, 2024. https://childrenshealthdefense.org/defender/uk-regulators-pregnant-women-no-covid-vaccines/.

# Index

**A**
ABMS. *See* American Board of Medical Specialties (ABMS)
ABOG. *See* American Board of Obstetrics & Gynecology (ABOG)
ACOG. *See* American College of Obstetricians and Gynecologists (ACOG)
adenoviral vaccines, 137
advertising, 57
AJOG. *See American Journal of Obstetrics and Gynecology* (AJOG)
Aldén, Markus, 45, 133
Ambros, Otto, 79
American Board of Medical Specialties (ABMS), 118–119, 123, 126

American Board of Obstetrics & Gynecology (ABOG), 58, 66, 85–86, 107–133, 140
American College of Obstetricians and Gynecologists (ACOG), 58–62, 66, 85, 127, 140
*American Journal of Obstetrics and Gynecology* (AJOG), 41, 43
antiphospholipid syndrome (APS), 144
APS. *See* antiphospholipid syndrome (APS)

**B**
Balbona, Eduardo, 144–145
Balletti, Brenda, 143
Banoun, Helen, 34
BBB. *See* blood-brain barrier (BBB)
Bhakdi, Sucharit, 131

Biden, Joe, 82, 137
biology, machine model of, 31–46
Biss, Kimberly, 148
Blaylock, Russell, 141
blood-brain barrier (BBB), 125, 141–142
Bowser, Muriel, 69
Bradford Hill Criteria, 23
Bridle, Byram, 37
Burkhardt, Arne, 131, 147

C
Camus, Albert, 47
Carlson, Tucker, 100
Caulfield, Tim, 64
causation, 23
CDC. *See* Centers for Disease Control (CDC)
Centers for Disease Control (CDC), 18, 44, 58–59, 62, 73, 108, 110–111, 114, 136, 148
Chemie Grünenthal, 77–79, 81
Colbert, Stephen, 70
Cole, Ryan, 42, 125
Collins, Frances, 67–68, 111
communication science, 59–61, 72–73
consent, informed, 38–43, 151–152
COVID-19 Community Corps, 59–62, 67–68, 71
Crick, Francis, 35
Cuomo, Andrew, 33, 134
cytotoxic cells, 140–141
Czech Republic, 167

D
Davison, Scott, 125
de Becker, Gavin, 107
de Garay, Maddie, 69–70
Delman, Meaney, 114
DES. *See* diethylstilbestrol (DES)
*Died Suddenly* (documentary), 16
diethylstilbestrol (DES), 21, 66, 74, 84–86
DNA, 35, 38, 46, 133, 151, 153. *See also* genome
Dobbs, Thomas, 112–113
Dodds, Edward Charles, 84
Drosten, Christian, 136
Drug Efficacy Amendment, 80, 82

E
Elledge, Kevin, 101, 103
Ellul, Jacques, 57
Emergency Use Authorization (EUA), 38–43, 135
employment, 100–106
endothelial cells, 141
EUA. *See* Emergency Use Authorization (EUA)
Evans, Harold, 75, 77–78

F
Facebook, 55
faith leaders, 67–69
Faiths4Vaccines, 67, 69
Fauci, Anthony, 32, 57, 139
FDA. *See* Food and Drug Administration (FDA)
Federation of State Medical Boards (FSMB), 108, 118–120, 123, 126
fetal death, 7, 21, 39, 91–92, 112, 125, 128, 130, 141. *See also* miscarriage; stillbirth
Fisher, Harry, 46–56

# Index

Food and Drug Administration (FDA), 18, 76, 79–81, 110, 114, 135, 137, 148–149
food supply, 85–86
Franklin, Rosalind, 35
FSMB. *See* Federation of State Medical Boards (FSMB)

## G
Geismar, Lee, 80
Gelsinger, Jesse, 34–37
Gelsinger, Paul, 34
gene therapy, 33–37, 127
genome, 35, 37, 44–46, 133, 151, 153. *See also* DNA
Golding, William, 100, 134
Grant, Kara, 110, 112

## H
Hanna, Nazeeh, 45
Hartman, Dan, 64
Health and Human Services (HHS), 18, 32, 58–62, 67, 70–71, 73, 101
Herbst, Arthur L., 84
HHS. *See* Health and Human Services (HHS)
Hick, Joseph, 167
Hodkinson, Roger, 3
Holmes, Elizabeth, 131
Houston Methodist Systems, 66
Huang, Jian-Pei, 146
Hudson, Cody, 143–153
hydroxychloroquine, 39, 44, 135, 153

## I
informed consent, 38–43, 151–152
"injure-to-kill" ratio, 21

iodine, 111
ivermectin, 39, 153

## J
Jackson, Brook, 17
James, Jason, 73
J. Gladstone Institute, 42–43
Jiro, Oyama, 80
job termination, 100–106
Johnson, Martin, 75
Johnson, Ron, 100

## K
Kefauver-Harris Amendment, 80, 82
Kelsey, Frances, 79–81, 153
Kennedy, John F., 75, 80–82
Kennedy, Robert F., Jr., 120, 168
killer T-cells, 140–141
Kirsch, Steve, 138
Klein, Naomi, 93–94
Kolletschka, Jakob, 156

## L
*Lancet,* 44, 112, 131, 135
Linard, Christian, 167
Lin study, 41–43
lipid nanoparticles (LNPs), 37, 125, 133, 141–143, 147
LNPs. *See* lipid nanoparticles (LNPs)
lupus, 144–145

## M
machine model, of biology, 31–46
Martin, Allen, 64–65
Martin, David, 159
Martin, Trista, 64–65
Mattis, James, 131

McClennon, Mary, 102
McCullough, Peter, 65, 134–135
McKernan, Kevin, 150–151
Mehra, Mandeep, 44
mifepristone, 21
miscarriage, 7, 18, 29–30, 40, 44, 47–48, 50–51, 84–85, 88–89, 91–92, 111, 113, 125, 145, 148. *See also* fetal death; stillbirth
misinformation, 108–109, 117, 121, 123
Moderna, 33, 56, 136, 150–151, 162
mRNA, 37, 41–46, 85–86, 127, 133, 136, 139–142, 151–153
Mückter, Heinrich, 79
Murdoch, Robert, 131
Murthy, Vivek, 67
Musk, Elon, 24
"my body, my choice," 93–94
myeloid cells, 140–141

## N

National Institute of Child Health and Human Development (NICHD), 42
Nazis, 79
*New England Journal of Medicine,* 44, 114, 130
NICHD. *See* National Institute of Child Health and Human Development (NICHD)

## O

"Obsolete Man, The" *(Twilight Zone),* 97–98
Oelrich, Stefan, 31

Operation Warp Speed, 134
Orwell, George, 97

## P

Parotto, Tiffany, 148
PEG. *See* polyethylene glycol (PEG)
Pfizer, 17–21, 23–24, 37–38, 40–41, 44, 48, 54–55, 57, 64, 69–70, 73, 88, 93, 105, 107, 111, 117, 119, 130, 136, 139–143, 147, 149–153, 168
*Pfizer Documents Analysis Report,* 40
polyethylene glycol (PEG), 142
Popper, Karl, 123–124
propaganda, 6, 32–33, 57, 62–63, 65, 70–72

## R

Ramirez, Earnest, 64
Rancourt, Denis, 138, 167
Raszek, Mikolaj, 46
Redfield, Robert, 168
Rinde, Meir, 36–37
Rose, Jessica, 138
rotavirus vaccine, 23
Rubin, Eric, 130

## S

*Sage's Newsletter,* 29
scientism, 35
Semmelweis, Ignaz, 105–106, 124, 155–158
Serling, Rod, 97
Sharav, Vera, 165
Shimabukuro, Tom, 44, 113–114
Siri, Aaron, 18, 40
SLE. *See* systemic lupus erythematosus (SLE)

# Index

SMFM. *See* Society for Maternal-Fetal Medicine (SMFM)
Society for Maternal-Fetal Medicine (SMFM), 58, 66, 85–86, 126–127
Spears, Donna, 102
Spencer, Michelle, 11–17, 93
SSM Health System, 67, 101, 103–105, 107
stillbirth, 6–7, 15, 21, 85, 89, 125, 140, 145. *See also* fetal death; miscarriage
Stock, Harald, 77
Strohman, Richard, 31, 34–35
suicide, 55
Sundgren, Nicolas, 162
SV40, 151
Swann, John P., 80–81
swine-flu vaccine, 23
systemic lupus erythematosus (SLE), 144–145

## T
termination, job, 100–106
thalidomide, 21, 66, 74–81
TikTok, 47–49, 55
trigeminal neuralgia, 55
*Twilight Zone*, 97–98
Twitter. *See* X (Twitter)

## U
UCSF. *See* University of California, San Francisco (UCSF)
University of California, San Francisco (UCSF), 42–43

## V
vaccination, defined, 135–136
vaccine, defined, 135
Vaccine Adverse Event Reporting System (VAERS), 105, 107, 117, 126–132, 138, 147–148
vaccine induced immune thrombotic thrombocytopenia (VITT), 137
VAERS. *See* Vaccine Adverse Event Reporting System (VAERS)
Vaughan, Robert, 3
vitamin C, 111
vitamin D, 111
VITT. *See* vaccine induced immune thrombotic thrombocytopenia (VITT)

## W
Walensky, Rochelle, 11, 90–91
Watson, James, 35
Weber, Mark, 59–60
Wendel, George, 116–118
White, Victoria, 160–163
Wilson, James, 36
Winsten, Jay, 59–60
Wirtz, Hermann, 79
Wolf, Naomi, 10, 40, 73, 87, 93–94, 145–147, 149

## X
X (Twitter), 17, 24, 47, 51, 55

## Y
Yeats, W. B., 154
Youth4Vaccines, 69

## Z
Zients, Jeff, 67
zinc, 111